AF698424

Long Non-Coding RNA

Long Non-Coding RNA:

From Disease Biomarkers to Targeted Therapeutics

Edited by

Syed Mudasir Ahmad
and Nazir Ahmad Ganai

Cambridge Scholars Publishing

Long Non-Coding RNA:
From Disease Biomarkers to Targeted Therapeutics

Edited by Syed Mudasir Ahmad and Nazir Ahmad Ganai

This book first published 2020

Cambridge Scholars Publishing

Lady Stephenson Library, Newcastle upon Tyne, NE6 2PA, UK

British Library Cataloguing in Publication Data
A catalogue record for this book is available from the British Library

Copyright © 2020 by Syed Mudasir Ahmad, Nazir Ahmad Ganai and contributors

All rights for this book reserved. No part of this book may be reproduced, stored in a retrieval system, or transmitted, in any form or by any means, electronic, mechanical, photocopying, recording or otherwise, without the prior permission of the copyright owner.

ISBN (10): 1-5275-5821-5
ISBN (13): 978-1-5275-5821-2

This book is dedicated to

Dr. A. P. J. Abdul Kalam
(1931–2015)

The whole of science is nothing more than a refinement of everyday thinking.

—Albert Einstein (1879–1955)

TABLE OF CONTENTS

PREFACE

In RNA biology, there has been a profound shift in our understanding of the role of non-coding RNA molecules. They were once regarded as mere templates for generating proteins, but growing evidence is proving they are becoming as large and diverse as their protein coding mRNA counterparts. LncRNAs have been found to be involved in a range of developmental processes and diseases, but details of the mechanisms by which they act are still surprising. It was suggested they play emerging roles in chromatin states and epigenetic and post-transcriptional gene regulations. The aberrant expression of lncRNAs and changes in their structure are associated with a wide spectrum of diseases. Some lncRNAs play a fundamental role in cellular identity, cell development, and disease progression. They are now widely studied in diverse organisms to understand their function. Thus, understanding the role of lncRNAs has become fundamental in biology, keeping in mind the fact of their emerging profile. An increasing number of experimental studies are providing evidence that lncRNAs mediate disease pathogenesis, thereby challenging the concept that protein-coding genes are the sole contributors to the development of human diseases.

LncRNAs are regulatory non-coding RNAs that are greater than 200 nucleotides in length; they have emerged as another critical layer of gene regulation. Given the significant role of LncRNAs in bimolecular regulatory interactions within cells and their prospects as drug targets in numerous diseases, it becomes highly imperative to understand their functionalities in disease pathogenesis. With this fascination, this book focuses on various types of disease to illustrate the potential impacts of lncRNAs on diverse biological processes and their therapeutic significance. This book is primarily centered on a few fundamental biological questions: How are lncRNAs involved in disease pathogenesis? How do lncRNAs regulate gene expression and contribute to cellular identities in physiological and pathological contexts, and how can they be targeted as potential biomarkers in drug therapeutics? This book is thus essential reading for biologists as well as interested researchers and students working in the domain of RNA biology. The book contains nine chapters.

Chapter 1 by Peerzada Tajamul Mumtaz and associates introduces the theme of the book, beginning with biogenesis and the mechanism of action. This chapter lays the foundation for the book by describing the types of lncRNAs and their characteristic features.

Chapter 2 by Eleonora Leucci summarizes current knowledge of the role of lncRNAs in cancer, and their potential clinical applications as biomarkers and therapeutic targets. It also provides an insight into the role of lncRNAs as major regulators of cancer cell plasticity and drug resistance.

Chapter 3 by Insha Zahoor et al. aims to provide a recent and cohesive but comprehensive update on the involvement of lncRNAs in MS pathogenesis, which will indeed lay down a much needed foundation for exploiting their potential as biomarkers in the early diagnosis of this complex disease and in predicting disease activity and its course, serving as novel targets for the development of alternative therapeutic interventions for its complete cure.

Chapter 4 by Qamar Taban and associates discusses the regulatory role of lncRNAs in both innate and adaptive branches of immunity and also enlightens us about their role in hematopoietic development, myeloid differentiation, inflammation, and host defense against pathogens.

Chapter 5 by Aashiq Hussain Bhat and associates details the lncRNAs involved in gene expression regulation and as mediators for diagnosis and prognosis, and as potential therapeutics in leukemias, hepatocellular carcinoma, and breast cancer.

Chapter 6 by Khalid Bashir Dar and associates gives mechanistic insights into the role of lncRNAs in brain function and more particularly in the pathogenesis of neurological diseases like Alzheimer's disease, Parkinson's disease, Angelman syndrome, and amyotrophic lateral sclerosis.

Chapter 7 by Mudasir Nabi et al. gives a detailed introduction to non-coding RNAs with special reference to lncRNAs and their biological role in various cellular pathways that contribute to the pathogenesis of PCOS. It also gives an insight into the future prospective of lncRNAs as non-invasive biomarkers and potential therapeutic agents in PCOS.

Chapter 8 by Dr. Bilal et al. gives us a brief insight into the role of lncRNAs in heart-related disorders and cardiovascular diseases (CVDs).

Chapter 9 by Dr. Syed Mudasir Ahmad gives an idea about the regulatory role of long non-coding RNAs and their role in host–pathogen interactions in general.

It is hoped that readers will find each chapter stimulating and thought provoking, which will add new dimensions to future RNA research. As editors of the book, we are indebted to the authors for the time and energy they spent in shaping the book, and the advancement of knowledge in the field that it generates. We are highly thankful to Prof. Khalid Rehman Hakeem, King Abdulaziz University, Jeddah, Saudi Arabia, for encouraging us to write a book in the field of LncRNA research. We are also grateful to Prof. Riaz Ahmad Shah, Head of the Department of Animal Biotechnology, FVSc and AH SKUAST-K, India, for his valuable suggestions in completing this assignment. Finally, we extend equal appreciation to our associate editors, Peerzada Tajamul Mumtaz and Qamar Taban, who wholeheartedly supported us in compiling the book. Gratitude is expressed to all the authors who have contributed chapters to this book.

This book is a part of the National Agricultural Higher Education Project (NAHEP) of SKUAST, Kashmir, India.

Dr. Syed Mudasir Ahmad
Division of Animal Biotechnology
Faculty of Veterinary Sciences
Husbandry, SKUAST-Kashmir, India

LIST OF ABBREVIATIONS

AD: Alzheimer's disease
AE–PCOS: Androgen Excess and PCOS Society
AGO2: Argonaute 2
Airn: antisense of insulin-like growth factor 2 non-coding RNA
ALL: acute lymphoblastic leukemia
ALS: amyotrophic lateral sclerosis
AML: acute myeloid leukemia
AMPK: AMP-activated protein kinase
ANRIL: antisense non-coding RNA in the INK4 locus
APCs: antigen-presenting cells
ASD: autism spectrum disorder
ASO: antisense oligonucleotides
BACE1-AS: B-secretase-1-AS lncRNA
BANCR: BRAF-activated non-coding RNA
BC: nrain cytoplasmic lncRNA
BDNF: brain derived neurotrophic factor
BDNF-AS: brain-derived neuro-trophic factor antisense RNA
BGL3: beta globin locus 3
Bvht: braveheart
CAA: chronic aplastic anemia
CAGE: cap analysis of gene expression
CARMEN: cardiac mesoderm enhancer-associated non-coding RNA
CCAT1: colon cancer–associated transcript 1
CCAT1-L: colorectal cancer associated transcript 1-long isoform
cDCs: classical DCs
CDKN2A: cyclin dependent kinase inhibitor 2A
cET: constrained EThyl
CETPH: chronic thromboembolic pulmonary hypertension
Chaer: cardiac hypertrophy-associated epigenetic regulator
CHAST: cardiac hypertrophy-associated transcript
CHRF: cardiac hypertrophy-related factor
CIS: clinically isolated syndrome
CNS: central nervous system
COX2: cyclooxygenase 2 gene
CPC: cardiac precursor cell

CTBP1-AS: C-Terminal binding protein 1 antisense
CTNNB1: oncogene β catenin
CVD: cardiovascular diseases
DDIT4: DNA damage inducible transcript 4
DYNLL1: dynein light chain 1
DYRK: dual-specificity tyrosine phosphorylation-regulated kinase
EBV: Epstein Barr virus
EMT: epithelial to mesenchymal transition
ERV: endogenous retroviral sequence
ESCs: embryonic stem cells
FAS-AS1: fas cell surface death receptor-antisense 1
FDA: food and drug administration
Fendrr: fetal-lethal non-coding developmental regulatory RNA
FGF19: fibroblast growth factor 19
FTLD: frontotemporal lobar degeneration
FUS/TLS: fused in sarcoma/translocated in sarcoma
GABA: g-aminobutyric acid
GAS5: growth arrest-specific transcript 5
GO: gene Ontology
H3K27ac: histone H3 Lys27 acetylation
HAR1: human accelerated region 1gene
HAT: histone acetyltransferase
HCC: hepatocellular Carcinomas
HD: Huntington's disease
HLA: human leukocyte antigen
hnRNPs: heterogeneous nuclear ribonucleoproteins
HOST2: human ovarian cancer-specific transcript
HOTAIR: HOX transcript antisense intergenic RNA
HOTAIRM1: HOXA transcript, antisense RNA myeloid-specific 1
HOTTIP: HOXA transcript at the distal tip
HOXC9: Homeobox C9
HSCs: hematopoietic stem cells
HSV: herpes simplex virus
HUVECs: human umbilical endothelial cell
IAV: influenza A
IGF2: insulin-like growth factor 2
IIM: idiopathic inflammatory myopathy
IL2RA: interleukin-2 receptor alpha
ISGs: interferon stimulated genes
JAK1: Janus kinase 1
KD: knock-down

KEGG: Kyoto Encyclopedia of Genes and Genomes analysis
KLF2: kruppel like factor 2
LINK-A: long intergenic NcRNA for Kinase Activation
Lmx1a: LIM homeobox transcription factor 1, alpha
LNA: locked nucleic acids
LncRNA: long non-coding RNAs
lncRNA-ATB: long non-coding RNA activated by TGF beta
LSD1: lysine specific demethylase 1
LTR: long terminal repeat
LUNAR1: Leukemia-induced non-coding activator rna-1
MALAT1: metastasis-associated lung adenocarcinoma transcript 1
MDRV: Muscovy Duck Reovirus
MeCP2: methyl CpG binding protein 2
MEG3: maternally expressed gene 3
MITF: microphtalmia-associated transcription factor
MLL: myeloid-lymphoid leukemia
Morrbid: myeloid RNA regulator of Bim-induced death
MS: multiple sclerosis
MSA: multiple system atrophy
MVIH: microvascular invasion
MYC: myelocytomatosis homologs
MyD: myeloid differentiation
NBAT-1: lncRNA neuroblastoma associated transcript-1
ncRNAs: non-coding RNAs
NEAT1: nuclear paraspeckles assembly transcript
NEIGHBOR: National Eye Institute Glaucoma Human Genetics Collaboration consortium
NeST: Nettoie Salmonella pas Theilers's
NFAT: Nuclear Factor of Activated T cells
NRAV: negative regulator of antiviral response
p53: tumor suppressor gene
PABP1: poly(A)-binding protein
PACER: p50-associated Cox2 extragenic RNA
PAN RNA: polyadenylated nuclear RNA
PANDAR: promoter of CDKN1A antisense DNA damage activated RNA
PCOS: Polycystic ovary syndrome
PD: Parkinson's disease
pDCs: Plasmacytoid DCs
PDX: patient derived xenograft
PGK1: Phosphoglycerate kinase 1
PINK1: PTEN -induced kinase 1

PMO: phosphorodiamidate morpholino oligomer
PNA: peptide nucleic acids
POAG: primary open-angle glaucoma
PPAR: proliferator-activated receptor
PPMO: peptide-conjugated PMO
PS: phosphorothioate
PTEN: phosphatase tensin homolog
PVT1: plasmacytoma variant translocation 1
RBPs: RNA-binding proteins
RISC: RNA-induced silencing complex
RMST: Rhabdomyosarcoma 2 associated transcript
RNA: ribonucleic acid
RNCR3: Retinal non-coding RNA 3
SAMMSON: survival-associated mitochondria melanoma-specific oncogenic ncRNA
SCA8: spinocerebellar ataxia type 8
SeV: Sendai virus
siRNA: short interfering RNA
SMILR: smooth muscle–induced lncRNA
Sorl1: sortilin-related receptor 1
SRA: steroid receptor RNA activator
STAT3: signal transducer and activator of transcription 3
SYNCRIP: synaptotagmin-binding cytoplasmic RNA interacting protein
TDP-43: transactive response DNA binding protein 43 kDa
TERT: telomerase reverse transcriptase
TGF-beta: transforming grow factor beta
THRIL: TNF and heterogeneous nuclear ribonucleoprotein L related immunoregulatory lincRNA
TLRs: toll-like Receptors
TSS: transcriptional start site
TUG1: taurine up-regulated gene 1
UBE3A: Ubiquitin protein ligase E3A
UCA1: urothelial cancer associated 1
Uchl1: ubiquitin carboxy-terminal hydrolase L1
XIST: X chromosome inactive-specific transcript
ZEB: zinc finger E-box-binding homeobox
Zeb2NAT: Zeb2 Natural antisense transcript

CHAPTER 1

FUNCTIONAL ASPECTS OF LONG NON-CODING RNAS: A REVIEW

PEERZADA TAJAMUL MUMTAZ,[1,2]
QAMAR TABAN,[1,3] SAMEENA ISMAIL,[4]
DIVYA SRIVASTAVA,[2]
MASHOOQ AHMAD DAR,[1]
RIAZ AHMAD SHAH,[1]
AND SYED MUDASIR AHMAD[*1]

[1] Division of Animal Biotechnology, Faculty of Veterinary Sciences and Animal Husbandry, Shuhama, SKUAST-K-190006, India
[2] Department of Biochemistry, School of Life Sciences, Jaipur National University, Jaipur, Rajasthan-302017, India
[3] Drug Therapeutics and Neurobiology Lab, Department of Biotechnology, School of Biological Sciences, University of Kashmir, Hazratbal, Srinagar, Jammu and Kashmir-190006, India
[4] Department of Biotechnology, Government Degree College Anantnag, Jammu and Kashmir-192101
*Corresponding author:
Syed Mudasir Ahmad,
Department of Animal Biotechnology, Faculty of Veterinary Sciences and Animal Husbandry, SKUAST-K
Tel: 9419514131; E-mail: mudasirbio@gmail.com

Abstract

Recent reports on RNA sequencing studies have revealed that most of the genome is transcribed, but very little of it encodes proteins, while the rest of the transcripts are known as non-coding transcripts. Long non-coding RNAs (lncRNAs) are non-coding transcripts longer than 200 nucleotides. Members of the non-coding genome include microRNA (miRNA), small regulatory RNAs, and other short RNAs. Most long non-coding RNAs (lncRNAs) are poorly annotated. Recent recognition of lncRNAs highlights their effects in many biological and pathological processes. LncRNAs are dysfunctional in a variety of human diseases varying from cancerous to non-cancerous diseases. Characterization of these lncRNA genes and their modes of action may allow their use for diagnosis, monitoring of progression, and targeted therapies in various diseases. In this review, we summarize the functional perspectives as well as the mechanism of action of lncRNAs.

Keywords: LncRNA, biogenesis, cellular localization, stability, functional utility

Introduction

As the complete transcriptional panorama of all organisms is more complex than was originally thought, the majority of the genomic sequence is arranged into a distinct array of protein-coding RNAs and non-coding RNAs (ncRNAs). Our index of genetic elements is brimming with long non-coding RNAs (lncRNAs), as in the genomic and transcriptomic view. Long non-coding RNAs are an enigmatic group of long RNA transcripts with no accepted protein-coding role. LncRNAs are usually found in every aspect of life; both the diversity and the size of non-coding RNA expression stocks rather than of protein-coding genes are associated with organismal complexity. Because of the diverse number of lncRNAs in humans, it is estimated that the number of lncRNAs is greater than the number of protein-coding genes. The total number of lncRNAs is continuously increasing with the advent of high-throughput next-generation sequencing (NGS) technology, the latest epigenomic technologies, and computational prediction techniques. Thus, understanding the role of lncRNAs has become fundamental in biology, keeping in view its emerging ranks. LncRNAs are regulatory non-coding RNAs that are greater than 200 nucleotides in length. On the basis of their roles, lncRNAs can be generally divided into three broad categories: the first category includes lncRNAs that are non-functional and likely to be the result of transcriptional noise; the second category includes lncRNAs for which the act of transcription alone is sufficient for their function but the transcript itself is not necessary; and the third category includes functional lncRNAs that act in cis and/or in trans ways. LncRNA transcripts have emerged as another critical layer of gene regulation.

Nomenclature of lncRNAs

LncRNAs are named after their function and expression pattern like XIST, HOTAIR. But this rule of nomenclature is inefficient due to the poor understanding of the functional utility of the majority of lncRNAs. To address such issues in the nomenclature of lncRNAs, the HUGO Gene Nomenclature Committee (HGNC) has drafted certain basic outlines to name unknown lncRNAs. The committee has decided to use the symbol "-AS" for anti-sense lncRNAs, as in "BOK-AS1" for "BOK antisense RNA 1," located in the opposite strand of BOK. Likewise, "-IT" is used for "intronic lncRNA," as in "MAGI2- IT1" for "MAGI2 intronic lncRNA transcript 1." The prefix "LINC" is followed by numbers used for intergenic lncRNAs (Volders et al. 2013; Wright and Bruford 2011).

LncRNA stability and turnover

After the study of the genome, it was found that the stability of about 7200 lncRNAs along with almost 20,000 protein-coding transcripts have a shorter turnover with a median of 3.5 hours compared with 5.1 hours for mRNAs. In this study, the half-lives of the lncRNAs ranged from 30 minutes to 48 hours and that of the protein-coding transcripts ranged from 2 hours to 12 hours (Niazi and Valadkhan 2012). This wide range of stability for lncRNA transcripts reflects the complexity of their metabolism and boundless functions (Clark et al. 2012; Wu et al. 2013). It was found that protein-coding genes with enzymatic and housekeeping functions were more stable as they did not involve any dynamic regulatory pathways. Since the regulatory elements are required for a short period, the proteins with regulatory activity have low stability. Correspondingly, lncRNAs involved in regulatory functions have shorter lifespans compared with housekeeping lncRNAs (Clark et al. 2012; Tani et al. 2012; Wu et al. 2013). A large number of lncRNAs are involved in the regulation of chromatin remodeling and gene expression in response to the external stimulus. The instability of lncRNAs with a regulatory function is paramount to their rapid dynamic regulation (Clark et al. 2012; Tani et al. 2012). Other factors such as subcellular localization, splicing, and genomic location also affects the stability of lncRNAs; this explains the greater stability of spliced lncRNAs than the single exon transcripts and lower stability of nuclear transcripts in line with their regulatory functions (Clark et al. 2012).

Protein coding potential of lncRNAs

To determine the protein-coding capability of lncRNAs, a number of experimental methods have been used (Ma et al. 2012). Further, different computational algorithms have been used to predict the protein-coding potential of lncRNAs and to determine the length of their ORFs. Another motive of these studies was to define the evolutionary conservation of lncRNA transcripts (Rinn and Chang 2012). These studies enlisted the shorter RNA transcripts (<300 nt) into the lncRNA group, since the ORF length of most of the eukaryotic protein-coding RNAs are longer than 300 nt or 100 amino acids. Studies have shown a significant difference between lncRNAs and mRNAs in their ORF length, with a median of 250 nt versus 1200 nt, respectively (Niazi and Valadkhan 2012), and also revealed that the start and stop codons in lncRNAs are distributed randomly among their genes (Ma et al. 2012). However, in eukaryotic cells some protein coding genes with an ORF shorter than 300 nt and lncRNAs with longer ORFs (>300 nt) have also been found (Ma et al. 2012). Therefore, other criteria like codon content, sequence context at −3 to +4 of the start codon, other protein domains, and so on were also employed to define the protein-coding capacity of these lncRNAs. The codon content of the lncRNA ORFs was found to be similar to the 3′UTRs of the protein-coding RNAs and to the intron codon alignment, indicating that lncRNAs and protein-coding RNAs have undergone evolutionary pressures through different pathways (Niazi and Valadkhan 2012). Further, it was shown that the ORFs of the longest lncRNAs lack the sequence context at −3 to +4 of the start codon indicating its translational efficiency (Niazi and Valadkhan 2012). LncRNA ORFs were also examined to detect known protein domains, which could be a sign for the protein-coding sequences (Volders et al. 2013). The details are reported on LNCipedia, a database for human long noncoding RNAs (http://www.lncipedia.org). Some other experimental methods such as ribosome profiling were used to identify the association of lncRNAs with ribosomes as well as their protein-coding potentials (Rinn and Chang 2012). Since all the susceptible translatable lncRNAs, which are associated with ribosomes, lack the evolutionary conservation of their proposed coding regions, such methods must be employed precisely (Rinn and Chang 2012). As a result, any interaction between ribosomes and lncRNAs does not necessarily result in translation. Further, it was shown that some of the ribosome-associated lncRNAs modulate the translation of protein-coding genes (Yoon et al. 2012). In conclusion, it is yet to be revealed whether lncRNAs possess any protein coding potential.

Genomic and cellular location of lncRNAs

LncRNAs can be found in either DNA strand or in both (Wang et al. 2011a). On the basis of genomic diversification with respect to protein-coding genes, lncRNAs can be classified into broad categories: for example, intergenic, intragenic, overlapping, and bidirectional lncRNA (figure 1). Sense lncRNAs are transcribed from loci, which overlap with a protein-coding gene; antisense lncRNAs are transcribed from loci that overlap with the antisense strand of the protein-coding gene; bidirectional lncRNAs are located on the antisense strand of a protein-coding gene whose start codon is <1000 base pairs away; intronic lncRNAs are transcribed from introns of protein-coding genes; and long intergenic noncoding RNAs (or intervening lncRNAs) loci are not in the proximity of a protein-coding gene at all (Da Sacco et al. 2012; Hauptman and Glavac 2013; Shahandeh 2013).

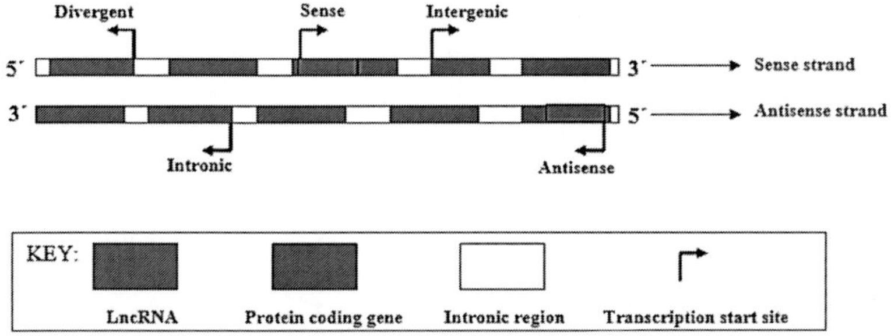

Figure 1. Different classes of long non-coding RNAs (lncRNAs). LncRNA genes can be classified into divergent, intronic, sense, antisense, and intergenic groups according to their location relative to the nearby protein-coding genes.

Additionally, these categories can be further classified into several subtypes. Prepared by the GENCODE project, among 14,880 lncRNAs, 9518 belong to intergenic lncRNAs and 5362 are from intragenic lncRNAs (Derrien et al. 2012). Intergenic lncRNAs are localized in the gene deserts, at least 5 kb away from protein-coding genes (Shahandeh 2013; Hauptman and Glavac 2013) and are also known as "large or long intergenic RNAs" (lincRNAs) (Knauss and Sun 2013). Enhancer RNA (eRNA) is a subclass of lincRNAs that is transcribed from enhancers and regulates the expression of its adjacent protein-coding genes (Wang and Chang 2011; Prensner and Chinnaiyan 2011). Beside, many other lincRNAs are transcribed from the gene-free chromosomal regions and regulate gene expression, both in cis and trans (Knauss and Sun 2013). Intragenic lncRNAs reside within a protein-coding gene and can be further sub-classified into four groups: (i) Sense-exonic lncRNAs, which overlap with one or more exons of protein-coding genes in the same strand. (ii) Antisense exonic lncRNAs, also known as "natural antisense transcripts" (NATs), which overlap with the antisense strand of a protein-coding gene and that can be transcribed from a host gene or from a different genomic locus with the same sequence complementarity (Da Sacco et al. 2012; Hauptman and Glavač 2013; Papait et al. 2013). NATs regulate gene expression through multiple mechanisms, for example through forming endo-siRNAs (Knauss and Sun 2013). (iii) Intronic sense and antisense lncRNAs are derived from an intron of a protein-coding transcript from the same or the opposite strand. (iv) 3′UTRs and 5′UTRs associated with RNAs or uaRNAs are derived from untranslated regions of the protein-coding transcripts (Nie et al. 2012; Wu et al. 2013). Overlapping lncRNA transcripts contain a protein-coding gene within an intron on the same strand (Derrien et al. 2012; Cheng et al. 2013). These lncRNAs may regulate the promoter of the overlapping genes. To date, 52 overlapping-sense lncRNAs have been identified in the human genome (Knauss and Sun 2013). Bidirectional or divergent lncRNAs are located in the opposite strand of an adjacent promoter and reside in fewer than approximately 1000 bp of a protein-coding gene (Wu et al. 2013; Knauss and Sun 2013; Shahandeh 2013). These lncRNAs usually have the same CpG island promoter as adjacent protein-coding genes and they also regulate the expression of their neighboring protein-coding genes (Sun et al. 2013). Figure 1 provides a schematic representation of the locus and transcription direction of lncRNAs. By identifying the subcellular localization of lncRNAs, scientists could be partially able to determine their functions (Lipovich et al. 2010). Analysis of the ENCODE RNAseq data for nucleus and cytoplasm revealed that the nucleus is highly enriched with lncRNA and that the lncRNAs are significantly located in the chromatin fraction indicating their regulatory roles (Derrien et al. 2012). Some lncRNAs are located only in the cytoplasm, and others in both the nucleus and the cytoplasm (Wang et al. 2011a). Thus, lncRNAs possess extensive functional capacity, especially in gene and chromatin regulation (Lipovich et al. 2010). Moreover, these findings verified Paul and Duerksen's research to some extent, in which it was reported that some parts of chromatin include RNAs twice as much as in DNAs. Figure 2 shows the cellular localization of lncRNAs.

Origin of lncRNAs

In eukaryotes, noncoding DNA has often been called "junk DNA," because with no evolutionary restriction, it experiences natural selection (Ponting et al. 2009). Each group, based on the differences between the nature of coding and non-protein-coding genes is differently disposed to evolutionary constraints. In contrast to protein-coding genes, non-coding genes are under secondary structure conservation (Gingeras 2007; Kutter et al. 2012). Since non-coding DNAs and protein-coding DNAs reveal different evolutionary pathways, they possess different conservations (Kaessmann 2010). Compared with protein-coding genes, we know little about the origin and evolution of lncRNAs. They show a low sequence conservation and rapid evolution among mammals (Kutter et al. 2012). In addition to this, various evolutionary assumptions have been considered for the origin of lncRNAs. The first hypothesis is the metamorphosis of the protein coding gene generated through a gene-duplication process (Ponting et al. 2009; Kaessmann 2010; Kapusta et al. 2013). During evolution, one copy of a protein-coding gene is exposed to accumulate mutations and eventually loses its protein-coding capacity. Then, a new functional lncRNA gene is designed that caters to some of the previous coding fractions including regulatory elements, splicing signals, exon sequences, and polyadenylation sequences (Ponting et al. 2009; Kaessmann 2010). An X inactive-specific transcript (XIST) is essential for sex-chromosome dosage compensation in mammals and was originated from the chicken protein-coding Lnx3 gene (Kaessmann 2010; Gribnau and Grootegoed 2012; Kapusta et al. 2013) and was then classified as a pseudo gene (Gerstein et al. 2007; Ponting et al. 2009). The XIST promoter region was originated from 5- UTR of Lnx3 exons 1 and 2. Further, exons 4 and 5 of the human XIST gene were generated from Lnx3 exons 4 and 11 (Elisaphenko et al. 2008). LncRNAs may be produced from segmental or whole-gene duplication of other lncRNA genes, which also include small and long non-coding RNAs (Ponting et al. 2009; Kapusta et al. 2013). The genomic examination of large homologous protein-coding gene families has shown that protein-coding gene duplication is a very common phenomenon; however, unlike protein-coding genes, there is very little evidence for the whole duplication of lncRNAs. This may be due to the rapid sequence divergence of lncRNAs (Ponting et al. 2009; Kaessmann 2010). Mouse nuclear-enriched abundant transcript 2 (Neat2) is an example of duplicated lncRNA that is paralogous to nonexonic sequences of the mouse genome (Ponting et al. 2009). Segmental gene duplication within ancRNA also seems to play a role in the generation of lncRNAs. Further, local and tandem duplications have been observed in the 5 regions of Kcnq1ot1 (Ponting et al. 2009; Kaessmann 2010). De novo generation is another possibility of the origins of lncRNAs (Kapusta et al. 2013). In this case, changes in the genomes such as chromosomal rearrangement and the generation of (proto-) splice sites and (proto-) promoters have transformed nonfunctional genomic sequences to functional lncRNAs (Kaessmann 2010). The last assumption considers the insertion of transposable elements (TEs) as the origin of lncRNAs (Ponting et al. 2009; Rinn and Chang 2012; Kapusta et al. 2013). In fact, TEs have given a lot, not only to the origin of new lncRNAs but also in terms of diversification and the regulation of vertebrate lncRNAs (Kapusta et al. 2013). Comparing lncRNAs with other genes like pseudo genes, small lncRNAs, and protein-coding genes, the majority of human lncRNAs have TE segments in their body. For instance, TEs are present within the internal part of exons, transcription start sites, polyadenylation (polyA) sites, or in the combination of all these parts. Studies have shown that 75% of human lncRNAs have at least one exon with partial TE origin. Most TE-based lncRNAs tend to have a few amounts of these sequences (Kapusta et al. 2013). Compared with other lncRNAs, intergenic lncRNAs are mostly covered by TEs (Rinn and Chang 2012). Four main classes of TEs are attributed to lncRNAs. However, LTR/ERV elements are the most abundant TE families among lncRNAs (Kapusta et al. 2013). TEs exert many important functions in lncRNA-related processes including the origin, diversification, and regulation of lncRNAs.

Role of TEs in lncRNA generation

The issue of generation of lncRNAs from transposable elements (TEs) is currently highly debated, thereby asking whether it is "lncRNA first" or "TE first": divergence or emergence? However, the scientific studies confirm both models (Kapusta et al. 2013). Some studies revealed that about 56.5% of TE-derived lncRNAs show some signatures of TEs to their last exons thereby supporting the "lncRNA-first" model. There is also some evidence that supports the "TE-first" model of generation of lncRNAs including the existence of lncRNAs that are found to mostly or entirely consist of TEs, as it seems to be improbable to consider the insertion of all these TEs into nonTE ancestral lncRNAs. Also there are some lncRNAs that are found within the start region and the first exon of transcripts derived from TEs. Human Gencode v13 catalogued about 4404 lncRNAs with TE-derived TSS and 2213 lncRNAs, whose first exons are covered by TEs and derived from primate-specific TE families (Kapusta et al. 2013). Findings supported both models and also unveiled that 40% of TEs-derived human lncRNAs are specific to primates and some are even restricted to hominoids (Kapusta et al. 2013). Therefore, these lncRNAs are considered to be young genes. Xist, which is developed from the Lnx3 pseudo gene and a set of diverse transposable elements, is one of the genes supporting both models. In this gene at least, a short fragment of the first exon, as well as exons 2, 3, 6, 7, 8, is reported to contain different types of TEs (Elisaphenko et al. 2008).

Role of TEs in the lineage-specific diversification of lncRNAs

Recently a large amount of evidence has evolved claiming a correlation between TEs' insertion in a regulatory region of genes and the emergence of lineage-specific lncRNAs. According to one study, about 36.3% of lncRNAs derived from TEs are found to be related to primate-specific lncRNAs (Kapusta et al. 2013). XIST, an lncRNA that evolved during the evolution of eutherians is found to have lineage-specific TEs like FLAM_C, lncRNA-RoR, and LTR7/HERVH, elements that cause varied expression of this lncRNA among certain species (Elisaphenko et al. 2008; Loh et al. 2011; Rinn and Chang 2012).

The role of TEs in the regulation of lncRNAs

Concordant TE insertion sites on lncRNA genes contribute to many aspects of lncRNA function and regulation. For instance, TEs contribute to transcription regulation of lncRNAs, their processing (e.g., splicing, alternative splicing, and polyadenylation) and the formation of lncRNAs' secondary structure formation (Kapusta et al. 2013; Hadjiargyrou and Delihas 2013). TEs that overlap with mapped DNAseI hypersensitive sites (DHS) indicate the active chromatin site and uncover the significant roles of TEs in lncRNA regulation. LTRs are revealed to be TEs that reside in the vicinity of the upstream region of lncRNAs. However, in protein-coding genes, SINEs are the TEs located in the proximal upstream region of lncRNAs (Kapusta et al. 2013). The distribution of these TEs in the upstream regulator sites of protein and noncoding genes is not random. TEs take part in cell type-specific regulations of specific lncRNAs, since different TEs-derived transcription regulatory motifs on DNA recruit different types of transcription factors. Consequently, this results in different spatio-temporal expression patterns (Hadjiargyrou and Delihas 2013; Kapusta et al. 2013). TEs located in lncRNA transcripts are involved in their functional regulation, such as acting as cis-regulators or trans-regulators of protein-coding gene expressions. ANRIL is one of these lncRNAs that regulates gene expression epigenetically. ANRIL binds to chromatin modifier factors and interacts with Alu supplementary motifs in promoters of ANRIL trans-regulating genes via its Alu motif (Holdt et al. 2013). One of the notable examples of gene expression transregulation is Alu-containing lncRNA originated and from chromosome 11. The Alu-containing lncRNAs bind to Alu sequences in 3′UTR of the target mRNAs forming specific secondary structures, recruiting Stau1 and subsequently UPF1. Their recruitments trigger target mRNA decay (Wang and Chang 2011; Hadjiargyrou and Delihas 2013).

Mechanism of action: lncRNAs as chromatin regulators

The number of lncRNAs with described functions is steadily increasing, and most of the reports revolve around their regulatory capacity. LncRNAs often function as important cis- and trans-acting modulators for the expression of protein-coding genes (Mao et al. 2011; Fejes-Toth et al. 2009). LncRNAs can mediate epigenetic modification by recruiting a chromatin-remodeling complex to a specific chromatin locus (figure 2A). At least 38% of the lncRNAs present in several tissues bind to the polycomb repressive complex 2 or the chromatin-modifying proteins, CoREST and SMCX (Fejes-Toth et al. 2009). Others bind to trithorax chromatin-activating complexes and/or activated chromatin (Kapranov et al. 2010). The well-characterized lncRNAs, ANRIL, XIST, HOTAIR, and KCNQ1OT1 are able to recruit epigenetic modifiers to specific loci for reprogramming the chromatin state. For example, KCNQ1OT1 binds to PRC2 and the methyl-transferase G9A (also known as EHMT2), whereas ANRIL binds to PRC1 and PRC2 (Lee et al. 2011; Mak et al. 2004). HOTAIR and other lncRNAs function as scaffolds (figure 2B) coordinating the targeting of specific repressive histone-modifying complexes to target loci (Lee et al. 2011). However, within this framework, the detailed mechanism of how specific DNA regions are targeted by lncRNAs remains unclear. An example of such principle approaches working together is the XIST locus that controls the X-chromosome dosage compensation. XIST, in many ways can be considered as a model example of lncRNA biology, and the dissection of its mechanism of action can serve as a guide for deeper and more thorough future lncRNA studies. Principally, X chromosome inactivation is due to the ablation of the DNA and the promoter of XIST (Mak et al. 2004; Brown et al. 1991); nevertheless, such models cannot undermine the effect of the local act of transcription.

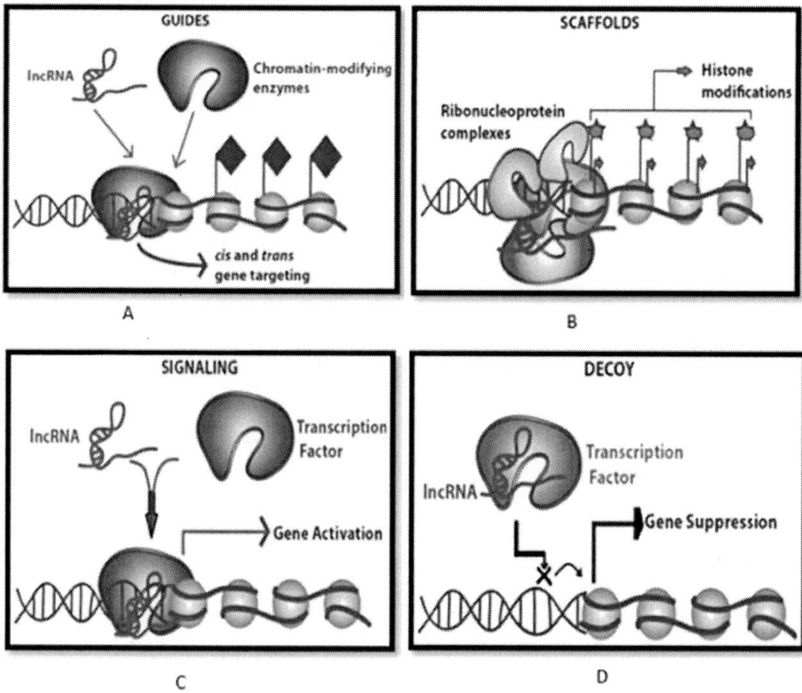

Figure 2. LncRNA-mediated transcriptional regulation. (Reproduced from our publication Bhat et al. 2016, *Noncoding RNA Research*).

Long non-coding RNAs (lncRNAs; red solid lines) regulate gene transcription through three main mechanisms: (A) Interaction with and recruitment of chromatin-modifying enzymes (e.g., histone methylases, acetylases, and deacetylases) to the target gene locus. Modulation of the chromatin state by these enzymes leads to activation or repression of local genes. (B) Interaction with other RNA-binding factors such as hnRNPs forms RNA-protein complexes (RNPs). RNPs can either promote transcription by recruiting key proteins to the target gene promoters or repress gene transcription by binding to existing gene repressors. (C) LncRNAs also have enhancer functions and help change the chromatin architecture and recruit transcriptional machinery proteins to the adjacent target gene locus to promote its transcription. (D) LncRNAs are also involved in the repression of certain pro-apoptotic genes, such as FAS and BIK, by acting as a decoy for the transcription factor (NF-YA).

Transcriptional regulation

Long non-coding RNAs act as co-factors to modify the activity of the transcriptional factor. For example, the ncRNA Evf2 is transcribed from the conserved distal enhancer and recruits the transcription factor (figure 2C) DlX2 to the same enhancer to induce expression of adjacent protein-coding genes (Zhao et al. 2008). Various developmental genes are regulated in a similar fashion by transcribing the enhancers in the cells in which they are active (Clemson et al. 1996). LncRNAs can modify RNA polymerase (RNAP) II activity by interplaying with the initiation complex to steer promoter choice; for example, it inhibits the binding of the transcriptional co-factor TFIID31 in human transcription of a ncRNA from an upstream region of the dihydrofolate reductase (DHFR) locus to form a triplex in the major promoter of DHFR. This could be a general mechanism for controlling the usage of the promoter, as thousands of these triplex structures exist in eukaryotic chromosomes (Sado et al. 2005). Long lncRNAs may also affect global changes by interacting with some basic components of the RNAP II-dependent transcription machinery. LncRNAs interacting with RNAP II machinery are primarily transcribed by RNAP III, thereby delinking their expression from the RNAP II-dependent transcription reaction they regulate. For example, in the transcription of Alu elements in response to heat shock, they bind tightly to RNAP II to rule out the formation of active pre-initiation complexes (Wan et al. 2008). Alu elements can independently mediate polymerase binding and repression by their respective domain interaction. Due to their abundance and distribution in the mammalian genome, these functional domains might have been co-operated into other ncRNAs during evolution (Edwards et al. 2007).

Post-transcriptional regulation

The ability of ncRNAs to identify complementary sequences allows some specific interactions capable of regulating post-transcriptional processing of mRNAs like capping, splicing, editing, transport, translation, degradation, and stability at various control sites. For example, MALAT1 affects alternate splicing by interacting with splicing factors. Another lncRNA, Gomafu/MIAT, which is localized to a nuclear domain and has a neuron-specific expression, may block spliceosome formation and affect mRNA splicing by sequestering splicing factor 1 (SF1) (He et al. 2014; Pandey et al. 2008). The NAT lncRNAs also play an essential role in regulating mRNA dynamics (figure 2D). About 61–72% of all transcribed regions possess lncRNAs in an antisense orientation (NATs) (Kanduri et al. 2006). Unlike classical NAT lncRNAs in the imprinting genes (Tsix, Air, HOTAIR, and Evf-2) that are responsible for recruiting repressor complexes like PRC2 to the target site, certain NATs result in RNA duplexes to inhibit cis-regulatory elements, leading to an alternate splicing pattern of the paired gene. For example, the Zeb2/Sip1 NAT complementary binds to the 5′ splice site of an intron in the 5′-UTR of the zinc finger Hox mRNA Zeb2, which plays a crucial role in epithelial-mesenchymal transition (EMT). Expression of the Zeb2 NAT upon EMT masks the splice site, hence preventing spliceosome action. Subsequently, the translation machinery can then recognize and bind to an internal ribosome entry site (IRES), resulting in more efficient Zeb2 translation.

The structural and functional diversity of lncRNAs, as well as their sheer number, make their characterization challenging: the NONCODE database, version 4, contains 95,135 human lncRNA transcripts (Mercer et al. 2013). The functional array of mechanisms of lncRNAs is diverse, ranging from scaffolds to decoys or signals (Xie et al 2014) that act either locally to regulate nearby genes in cis or remotely, that is, over one megabase away or on a different chromosome, in trans, to control the transcription of target genes (Klattenhoff et al. 2013). LncRNAs could directly silence or activate gene expression, or indirectly regulate the chromatin states of their target genes. Indeed, many lncRNAs are very much like mRNAs: they are transcribed by RNA polymerase II (Pol II) from genomic loci with similar chromatin states to mRNAs; they are often 5′-capped, spliced, and polyadenylated; and in most instances, they lack any biochemical distinction from mRNAs beside the absence of a translated ORF. As a class, lncRNAs run the gamut from mRNA-like to truly exotic, such as circular RNAs (He et al. 2013), lncRNAs spanning 100 kb (Bell et al. 2014), and lncRNAs with a restricted subnuclear localization (Wang et al. 2014a). When compared with mRNAs, lncRNAs are more enriched in the nucleus relative to the cytoplasm and within the nucleus; they occupy the chromatin fraction (17% of lncRNAs versus 15% of mRNAs are enriched in the nucleus, whereas 4% versus 26%, respectively, are enriched in the cytoplasm). This observation agrees with a report that lncRNA-read density is higher in the nuclear fraction. Nuclear lncRNAs can recruit sequestering proteins, such as transcription factors or histone-modifying enzymes, to modify the transcription from target loci; and cytoplasmic lncRNAs regulate protein localization, mRNA translation, and their stability in the cytoplasm. Many lncRNA species have very well-defined subcellular localizations, including XIST (on the inactive X), Gomafu (also known as MIAT; subnuclear domains), BORG (restricted to the nucleus), and GAS5 (exported to the cytoplasm). Several degradation pathways preferentially act on lncRNAs, including subsequent 5′-to-3′ exonuclear degradation via Xrn1, or by nonsense-mediated decay (Wang et al. 2014b)

LncRNAs are involved in numerous important biological phenomena such as imprinting genomic loci, shaping chromosome conformation, and allosterically regulating enzymatic activity and pathological mechanisms. The over-expression, deficiency, or mutation of lncRNA genes has been implicated in numerous diseases. By now, although the function of most lncRNAs is yet to be discovered, several lncRNAs have been clearly shown to play critical roles through their molecular mechanisms in various cellular processes, which include differentiation, growth, apoptosis, development, and tumorigenesis (Zangrando et al. 2014). Increasing evidence firmly supports the fact that lncRNAs may act as diagnostic markers or therapeutic targets in many diseases. For example, lncRNA H19 acts as a carcinogenic gene and was found to be involved in gastric cancer, colorectal cancer, and glioma. HOTAIR mediates epigenetic silencing of another HOX cluster on a different chromosome via binding PRC2. Its over-expression in breast cancer, hepatocellular carcinoma, colorectal cancer, and pancreatic cancer is associated with poor prognosis and tumor metastasis as well as invasion ability in gastrointestinal stromal tumors. It has been elucidated that there is a relationship between ANRIL and increased susceptibility to several types of cancer such as acute lymphoblastic leukemia, glioma, basal cell carcinoma, nasopharyngeal carcinoma, breast cancer, and plexiform neurofibromas (Ounzain et al. 2014). Another strong study proves that MALAT1 (metastasis-associated lung adenocarcinoma transcript 1) might be a screening biomarker for lung cancer, uterine endometrial stromal sarcoma, cervical cancer, and hepatocellular carcinoma because of its association with metastasis. It promotes the motility of lung cancer cells through transcriptional or post-transcriptional regulation of motility-related genes. The lncRNA SLINKY has been found to be a robust prognostic biomarker in clear cell renal cell carcinomas (ccRCC), where it has been shown to function together with HNRNPK in cancer cell proliferation (Yang et al. 2014). A study also suggested that the dysregulation of lncRNA SNHG6 is highly cancer-type specific (Kumarswamy et al. 2014).

The lncRNA BACE1AS has been reported to play a vital role in the etiology of Alzheimer's disease (AD). Various lncRNAs including the Gas5 transcript is involved in host-pathogen interactions (Vausort et al. 2014). Although research on lncRNAs in endocrine diseases remains limited, the important roles of several genes in the metabolism and endocrine system have been reported, such as the PTEN-induced gene PINK1, which has been shown to be associated with diabetic status. The

FADS was found to be regulated by the dietary fat content. Researchers showed that the abnormal expression of MIAT (myocardial infarction associated transcript) is involved in cell proliferation, apoptosis, and migration in many diseases, such as myocardial infarction, microvascular dysfunction, and diabetes. Studies have also explored lncRNAs as biomarkers for the context of cardiovascular diseases. Recently, a study reported lncRNAs in the plasma of patients with myocardial infarction (MI) (Michalik et al. 2014). This list of characterized lncRNAs continues to grow as researchers go deeper and deeper into this amazing repository of lncRNAs.

LncRNAs are endorsed to be potential biomarkers in the diagnosis and prognosis of bladder cancer, prostate cancer, gastric cancer, pancreatic cancer, breast cancer, and many other cancer types as they possess high specificity, sensitivity, non-invasive characteristics, and easy detection in the tissues, serum, plasma, urine, and saliva. LncRNA PCA3 is now routinely used in clinics for the diagnosis of prostate cancer. Some lncRNAs have also been used as an adjunct to improve the specificity and sensitivity of existing biomarkers. Molecular tools such as RNA-seq, RNA-FISH, ic-SHAPE, and quantitative real-time PCR have been used to examine the potential of lncRNAs. Single nucleotide polymorphisms (SNPs) in lncRNAs can also be used as a predictor of cancer risk. Although ongoing studies continue to unravel the underlying mechanism, some lncRNAs have been used as therapeutic targets for the selective killing of cancer cells in patients. Thus, conclusively, lncRNAs have a great potential for convenient and minimally invasive diagnostic/prognostic markers, and also as therapeutic targets.

Acknowledgement

We offer sincere gratitude to the Council of Scientific and Industrial Research (CSIR), Government of India, for financial support in the form of Junior and Senior Research Fellowships.

Conflict of interest

The authors declare that they have no conflicts of interest with respect to research, authorship, and publication of this book chapter.

Copyright and permission statement

We confirm that the materials included in this chapter do not violate copyright laws. Where relevant, appropriate permissions have been obtained from the original copyright holder(s). All original sources have been appropriately acknowledged and/or referenced.

Funding

No funding was received for this project.

References

Bell, R. D., Long, X., Lin, M., Bergmann, J. H., Nanda, V., Cowan, S. L., . . . & Miano, J. M. (2014). Identification and initial functional characterization of a human vascular cell-enriched long noncoding RNA. *Arteriosclerosis, thrombosis, and vascular biology*, *34*(6), 1249–59.

Brown, C. J., Lafreniere, R. G., Powers, V. E., Sebastio, G., Ballabio, A., Pettigrew, A. L., . . . & Willard, H. F. (1991). Localization of the X inactivation centre on the human X chromosome in Xq13. *Nature*, *349*(6304), 82.

Chen, G., Wang, Z., Wang, D., Qiu, C., Liu, M., Chen, X., . . . & Cui, Q. (2012). LncRNADisease: a database for long-non-coding RNA-associated diseases. *Nucleic acids research*, *41*(D1), D983–D986.

Clark, M. B., Johnston, R. L., Inostroza-Ponta, M., Fox, A. H., Fortini, E., Moscato, P., . . . & Mattick, J. S. (2012). Genome-wide analysis of long noncoding RNA stability. *Genome research*, *22*(5), 885–898.

Clemson, C. M., McNeil, J. A., Willard, H. F., & Lawrence, J. B. (1996). XIST RNA paints the inactive X chromosome at interphase: evidence for a novel RNA involved in nuclear/chromosome structure. *Journal of cell biology*, 132(3), 259–75.

Sacco, L. D., Baldassarre, A., & Masotti, A. (2012). Bioinformatics tools and novel challenges in long non-coding RNAs (lncRNAs) functional analysis. *International journal of molecular sciences*, *13*(1), 97–114.

Derrien, T., Johnson, R., Bussotti, G., Tanzer, A., Djebali, S., Tilgner, H., . . . & Lagarde, J. (2012). The GENCODE v7 catalog of human long noncoding RNAs: analysis of their gene structure, evolution, and expression. *Genome research*, *22*(9), 1775–89.

Edwards, C. A., & Ferguson-Smith, A. C. (2007). Mechanisms regulating imprinted genes in clusters. *Current opinion in cell biology*, *19*(3), 281–89.

Elisaphenko, E. A., Kolesnikov, N. N., Shevchenko, A. I., Rogozin, I. B., Nesterova, T. B., Brockdorff, N., & Zakian, S. M. (2008). A dual origin of the Xist gene from a protein-coding gene and a set of transposable elements. *PloS one*, *3*(6), e2521.

Fejes-Toth, K., Sotirova, V., Sachidanandam, R., Assaf, G., Hannon, G. J., Kapranov, P., . . . & Gingeras, T. R. (2009). Post-transcriptional processing generates a diversity of 5′-modified long and short RNAs: Affymetrix/Cold Spring Harbor Laboratory ENCODE Transcriptome Project. *Nature*, *457*(7232), 1028.

Gerstein, M. B., Bruce, C., Rozowsky, J. S., Zheng, D., Du, J., Korbel, J. O., . . . & Snyder, M. (2007). What is a gene, post-ENCODE? History and updated definition. *Genome research*, *17*(6), 669–81.

Gingeras, T. R. (2007). Origin of phenotypes: genes and transcripts. *Genome research*, *17*(6), 682–90.

Gribnau, J., & Grootegoed, J. A. (2012). Origin and evolution of X chromosome inactivation. *Current opinion in cell biology*, *24*(3), 397–404.

Hadjiargyrou, M., & Delihas, N. (2013). The intertwining of transposable elements and non-coding RNAs. *International journal of molecular sciences*, *14*(7), 13307–28.

Hauptman, N., & Glavač, D. (2013). Long non-coding RNA in cancer. *International journal of molecular sciences*, *14*(3), 4655–69.

He, S., Gu, W., Li, Y., & Zhu, H. (2013). ANRIL/CDKN2B-AS shows two-stage clade-specific evolution and becomes conserved after transposon insertions in simians. *BMC evolutionary biology*, *13*(1), 247.

He, Y., Meng, X. M., Huang, C., Wu, B. M., Zhang, L., Lv, X. W., & Li, J. (2014). Long noncoding RNAs: novel insights into hepatocelluar carcinoma. *Cancer letters*, *344*(1), 20–27.

Holdt, L. M., Hoffmann, S., Sass, K., Langenberger, D., Scholz, M., Krohn, K., . . . & Gielen, S. (2013). Alu elements in ANRIL non-coding RNA at chromosome 9p21 modulate atherogenic cell functions through trans-regulation of gene networks. *PLoS genetics*, *9*(7), e1003588.

Kaessmann, H. (2010). Origins, evolution, and phenotypic impact of new genes. *Genome research*, *20*(10), 1313–26.

Kanduri, C., Thakur, N., & Pandey, R. R. (2006). The length of the transcript encoded from the Kcnq1ot1 antisense promoter determines the degree of silencing. *EMBO journal*, *25*(10), 2096–106.

Kapranov, P., St Laurent, G., Raz, T., Ozsolak, F., Reynolds, C. P., Sorensen, P. H., . . . & Triche, T. J. (2010). The majority of total nuclear-encoded non-ribosomal RNA in a human cell is "dark matter" un-annotated RNA. *BMC biology*, *8*(1), 149.

Kapusta, A., Kronenberg, Z., Lynch, V. J., Zhuo, X., Ramsay, L., Bourque, G., . . . & Feschotte, C. (2013). Transposable elements are major contributors to the origin, diversification, and regulation of vertebrate long noncoding RNAs. *PLoS genetics*, *9*(4), e1003470.

Klattenhoff, C. A., Scheuermann, J. C., Surface, L. E., Bradley, R. K., Fields, P. A., Steinhauser, M. L., . . . & Abo, R. (2013). Braveheart, a long noncoding RNA required for cardiovascular lineage commitment. *Cell*, *152*(3), 570–83.

Knauss, J. L., & Sun, T. (2013). Regulatory mechanisms of long noncoding RNAs in vertebrate central nervous system development and function. *Neuroscience*, *235*, 200–214.

Kumarswamy, R., Bauters, C., Volkmann, I., Maury, F., Fetisch, J., Holzmann, A., . . . & Thum, T. (2014). Circulating long noncoding RNA, LIPCAR, predicts survival in patients with heart failure. *Circulation research*, *114*(10), 1569–75.

Kutter, C., Watt, S., Stefflova, K., Wilson, M. D., Goncalves, A., Ponting, C. P., . . . & Marques, A. C. (2012). Rapid turnover of long noncoding RNAs and the evolution of gene expression. *PLoS genetics*, *8*(7), e1002841.

Lee, J. T. (2011). Gracefully ageing at 50, X-chromosome inactivation becomes a paradigm for RNA and chromatin control. *Nature reviews Molecular cell biology*, *12*(12), 815.

Lipovich, L., Johnson, R., & Lin, C. Y. (2010). MacroRNA underdogs in a microRNA world: evolutionary, regulatory, and biomedical significance of mammalian long non-protein-coding RNA. *Biochimica et Biophysica Acta (BBA)-Gene Regulatory Mechanisms*, *1799*(9), 597–615.

Loh, Y. H., Yang, L., Yang, J. C., Li, H., Collins, J. J., & Daley, G. Q. (2011). Genomic approaches to deconstruct pluripotency. *Annual review of genomics and human genetics*, *12*, 165–85.

Ma, H., Hao, Y., Dong, X., Gong, Q., Chen, J., Zhang, J., & Tian, W. (2012). Molecular mechanisms and function prediction of long noncoding RNA. *Scientific World Journal*, *2012*.

Mak, W., Nesterova, T. B., de Napoles, M., Appanah, R., Yamanaka, S., Otte, A. P., & Brockdorff, N. (2004). Reactivation of the paternal X chromosome in early mouse embryos. *Science*, *303*(5658), 666–69.

Mao, Y. S., Sunwoo, H., Zhang, B., & Spector, D. L. (2011). Direct visualization of the co-transcriptional assembly of a nuclear body by noncoding RNAs. *Nature cell biology*, *13*(1), 95.

Mercer, T. R., & Mattick, J. S. (2013). Structure and function of long noncoding RNAs in epigenetic regulation. *Nature structural & molecular biology*, *20*(3), 300.

Michalik, K. M., You, X., Manavski, Y., Doddaballapur, A., Zörnig, M., Braun, T., . . . & Boon, R. A. (2014). Long noncoding RNA MALAT1 regulates endothelial cell function and vessel growth. *Circulation research*, *114*(9), 1389–97.

Niazi, F., & Valadkhan, S. (2012). Computational analysis of functional long noncoding RNAs reveals lack of peptide-coding capacity and parallels with 3′ UTRs. *Rna*, *18*(4), 825–43.

Nie, L., Wu, H. J., Hsu, J. M., Chang, S. S., LaBaff, A. M., Li, C. W., . . . & Hung, M. C. (2012). Long non-coding RNAs: versatile master regulators of gene expression and crucial players in cancer. *American journal of translational research*, *4*(2), 127.

Ounzain, S., Micheletti, R., Beckmann, T., Schroen, B., Alexanian, M., Pezzuto, I., . . . & Dauvillier, J. (2014). Genome-wide profiling of the cardiac transcriptome after myocardial infarction identifies novel heart-specific long non-coding RNAs. *European heart journal*, *36*(6), 353–68.

Pandey, R. R., Mondal, T., Mohammad, F., Enroth, S., Redrup, L., Komorowski, J., . . . & Kanduri, C. (2008). Kcnq1ot1 antisense noncoding RNA mediates lineage-specific transcriptional silencing through chromatin-level regulation. *Molecular cell*, *32*(2), 232–46.

Papait, R., Kunderfranco, P., Stirparo, G. G., Latronico, M. V., & Condorelli, G. (2013). Long noncoding RNA: a new player of heart failure? *Journal of cardiovascular translational research*, *6*(6), 876–83.

Ponting, C. P., Oliver, P. L., & Reik, W. (2009). Evolution and functions of long noncoding RNAs. *Cell*, *136*(4), 629–41.

Prensner, J. R., & Chinnaiyan, A. M. (2011). The emergence of lncRNAs in cancer biology. *Cancer discovery*, *1*(5), 391–407.

Rinn, J. L., & Chang, H. Y. (2012). Genome regulation by long noncoding RNAs. *Annual review of biochemistry*, *81*, 145–66.

Sado, T., Hoki, Y., & Sasaki, H. (2005). Tsix silences Xist through modification of chromatin structure. *Developmental cell*, *9*(1), 159–65.

Shahandeh, A. (2013). Molecular mechanisms of oncogenic long non-coding RNAs. *Biosci. Res*, *10*, 38–54.

Tani, H., Mizutani, R., Salam, K. A., Tano, K., Ijiri, K., Wakamatsu, A., . . . & Akimitsu, N. (2012). Genome-wide determination of RNA stability reveals hundreds of short-lived noncoding transcripts in mammals. *Genome research*, *22*(5), 947–56.

Vausort, M., Wagner, D. R., & Devaux, Y. (2014). Long noncoding RNAs in patients with acute myocardial infarction. *Circulation research*, *115*(7), 668–77.

Volders, P. J., Helsens, K., Wang, X., Menten, B., Martens, L., Gevaert, K., . . . & Mestdagh, P. (2012). LNCipedia: a database for annotated human lncRNA transcript sequences and structures. *Nucleic acids research*, *41*(D1), D246–D251.

Wan, L. B., & Bartolomei, M. S. (2008). Regulation of imprinting in clusters: noncoding RNAs versus insulators. *Advances in genetics*, *61*, 207–23.

Wang, K., Liu, F., Zhou, L. Y., Long, B., Yuan, S. M., Wang, Y., . . . & Li, P. F. (2014). The long noncoding RNA CHRF regulates cardiac hypertrophy by targeting miR-489. *Circulation research*, *114*(9), 1377–88.

Wang, K., Long, B., Zhou, L. Y., Liu, F., Zhou, Q. Y., Liu, C. Y., . . . & Li, P. F. (2014). CARL lncRNA inhibits anoxia-induced mitochondrial fission and apoptosis in cardiomyocytes by impairing miR-539-dependent PHB2 downregulation. *Nature communications*, *5*, 3596.

Wang, K. C., & Chang, H. Y. (2011). Molecular mechanisms of long noncoding RNAs. *Molecular cell*, *43*(6), 904–14.

Wang, X., Song, X., Glass, C. K., & Rosenfeld, M. G. (2011). The long arm of long noncoding RNAs: roles as sensors regulating gene transcriptional programs. *Cold Spring Harbor perspectives in biology*, *3*(1), a003756.

Wright, M. W., & Bruford, E. A. (2011). Naming "junk": human non-protein coding RNA (ncRNA) gene nomenclature. *Human genomics*, *5*(2), 90.

Wu, P., Zuo, X., Deng, H., Liu, X., Liu, L., & Ji, A. (2013). Roles of long noncoding RNAs in brain development, functional diversification and neurodegenerative diseases. *Brain research bulletin*, *97*, 69–80.

Xie, C., Yuan, J., Li, H., Li, M., Zhao, G., Bu, D., . . . & Zhao, Y. (2013). NONCODEv4: exploring the world of long non-coding RNA genes. *Nucleic acids research*, *42*(D1), D98–D103.

Yang, K. C., Yamada, K. A., Patel, A. Y., Topkara, V. K., George, I., Cheema, F. H., . . . & Nerbonne, J. M. (2014). Deep RNA sequencing reveals dynamic regulation of myocardial noncoding RNAs in failing human heart and remodeling with mechanical circulatory support. *Circulation*, *129*(9), 1009–21.

Yoon, J. H., Abdelmohsen, K., Srikantan, S., Yang, X., Martindale, J. L., De, S., . . . & Gorospe, M. (2012). LincRNA-p21 suppresses target mRNA translation. *Molecular cell*, *47*(4), 648–55.

Zangrando, J., Zhang, L., Vausort, M., Maskali, F., Marie, P. Y., Wagner, D. R., & Devaux, Y. (2014). Identification of candidate long non-coding RNAs in response to myocardial infarction. *BMC genomics*, *15*(1), 460.

Zhao, J., Sun, B. K., Erwin, J. A., Song, J. J., & Lee, J. T. (2008). Polycomb proteins targeted by a short repeat RNA to the mouse X chromosome. *Science*, *322*(5902), 750–56.

CHAPTER 2

CANCER DEVELOPMENT AND THERAPY RESISTANCE: SPOTLIGHTS ON THE DARK SIDE OF THE GENOME

ELEONORA LEUCCI*[1]

[1] Department of Oncology, Laboratory of RNA Cancer Biology, LKI Leuven Cancer Institute, KU Leuven-University of Leuven, Herestraat 49, 3000 Leuven, Belgium
*Corresponding author:
Eleonora Leucci, Assistant Professor
Laboratory for RNA Cancer Biology, Department of Oncology, KU Leuven, LEUVEN
Tel: +32 471721179; E-mail: eleonora.leucci@kuleuven.be

Abstract

Cancer research has so far focused on genetic alterations in protein-coding genes. However, mounting evidence underlines the importance of epigenetic and post-transcriptional events in cancer progression and therapy resistance. Moreover, recent genome-wide studies show that disease-causing mutations and chromosome rearrangements often span areas of the genome that do not contain any known protein-coding gene. This finding is not surprising, considering that even though the vast majority of the human genome is transcribed, only a minor portion (accounting for less than 2%) encodes for proteins. Among the various classes of transcribed RNAs, long non-coding RNAs are attractive biomarkers and therapeutic targets due to their disease- and stage-restricted expression.

In addition, by taking part in all the major epigenetic and post-transcriptional programs in the cell, long non-coding RNAs are key regulators of stress responses and therefore important non-genetic players in cancer development and progression. Here I discuss the role of lncRNAs in cancer and their promises and pitfalls as biomarkers and therapeutic targets.

Keywords: lncRNA, cancer, biomarkers, therapy resistance, personalized medicine

Introduction

Despite the latest advancement in treatment, cancer is still the second leading cause of death worldwide (source: World Health Organization). One major reason for this is that clinical responses to anticancer drugs are either transient or limited to restricted subsets of patients because of intrinsic or acquired resistances (Holohan et al. 2013). In addition, most cancer research so far has focused on protein-coding genes that account for only 2% of the human genome, in a scenario where most of the genome is instead pervasively transcribed into several non-coding RNA species (FANTOM Consortium 2005).

Among those, long non-coding RNAs (lncRNAs) represent a heterogeneous class of transcripts longer than 200 nucleotides and lacking evidence for coding potential (Banfai et al. 2012; Guttman et al. 2013). They are generally transcribed by RNA polymerase II and therefore they are often spliced, capped, and polyadenylated (Quinn and Chang 2016). LncRNAs account for a large portion of so-called junk DNA, the dark matter of the genome, that has recently been shown to host the vast majority of cancer-related recurrent somatic mutations (Melton et al. 2015), single-nucleotide polymorphisms (Cheetham et al. 2013), and copy number alterations (Beroukhim et al. 2010). Mounting evidence supports a link between lncRNA deregulation and cancer development and progression, including resistance to therapy.

Analysis of conservation shows that lncRNAs have been extensively remodeled during evolution by transposable elements and therefore share only short areas of conserved sequences often biased towards the 5′- end of the transcript (Hezroni et al. 2015). The lack of conservation has halted the development of genetic *in vivo* models, which could univocally prove the implication of lncRNAs in cancer development and/or progression. Those models are available only for a handful of well-conserved transcripts (Adriaens et al. 2016; Arun et al. 2016; Yildirim et al. 2013; Mello et al. 2017; Hosono et al. 2017), while in the case of primate-specific lncRNAs, patient-derived xenografts (PDX) models are also emerging as a valid alternative (Leucci et al. 2016). PDXs are also an essential tool to test lncRNA-based drugs in preclinical settings alone or in combination with the standard of care.

It is worth noting that even though their sequences are not conserved, many lncRNAs actually have evolutionary conserved functions

(Ulitsky et al. 2011) and their promoters have unusually high sequence conservation, in contrast to the transcripts to which they give birth (FANTOM consortium 2005).

In terms of molecular mechanisms, it looks like lncRNAs have evolved highly diverse ways of hijacking the cellular machinery, ranging from recruitment of chromatin remodeling enzymes (Yap et al. 2010) to sponging of microRNAs and proteins (Poliseno et al. 2010; Kino T et al. 2010) (figure 1). It is therefore unsurprising to discover that they have been implicated in virtually all physiological and pathological processes in the cell (Huarte 2015; Lorenzen and Thum 2016; Fatica and Bozzoni 2014).

Despite the heterogeneity in molecular mechanism, however, it is accepted that lncRNAs are mostly involved in the regulation of epigenetic and post-transcriptional events (figure 1).

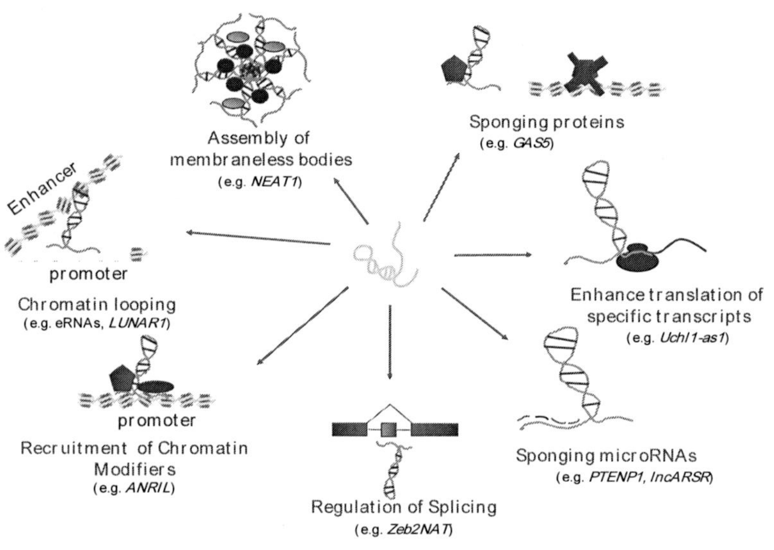

Figure 1. Archetypes of lncRNA functions in the cell.

LncRNAs participate in several epigenetics and post-transcriptional events in the cell. NEAT1 (Clemson et al. 2009) participates in the formation of membraneless bodies in the nucleus and is able to trap proteins and nucleic acids. GAS5 acts as a decoy for the glucocorticoid receptor by mimicking the glucocorticoid response elements (Kino et al. 2010). Uchl1-as1 overlaps with the 5′ of Uchl1 and runs in an antisense direction. By base-pairing to the sense strand, it enhances the recruitment to the ribosome and the translation of its sense (Carrieri et al. 2012). PTENP1 is a pseudogene of the tumor suppressor PTEN and therefore shares with it its sequence. When expressed PTENP1 sponges PTEN-targeting microRNAs thus stabilizing PTEN mRNA (Poliseno et al. 2010). A similar mechanism is used by lncARSRs that sponge miR-34 and miR-449 from targeting *AXL* and *MET* mRNA (Qu et al. 2016). The Zeb2NAT sequence overlaps the 5′-splice site of a large intron in Zeb2 mRNA. By base-paring to the sequence, Zeb2NAT prevents its splicing (Beltran et al. 2007). ANRIL, an antisense to the tumor-suppressor locus INK4b/ARF/INK4a, recruits CBX7 (chromobox 7) to the locus thus causing H3K27 (Histone 3 Lysine 27) methylation and the silencing of the locus (Yap et al. 2010). eRNA produced by transcription from the enhancer regions favors chromatin looping with the promoters (Ørom and Shiekhattar 2013). *LUNAR1* is able to control the expression of the *IGF1R* gene through chromosomal looping (Trimarchi et al. 2014).

Cell plasticity and inter- and intra-tumor heterogeneity are two major sources of therapy resistance (Holohan et al. 2013). Although most of the work on drug resistance has focused on pre-existing or acquired mutations, it is becoming increasingly clear that non-genetic events preceding the acquisition of resistance also play a major role (Sharma et al. 2010). Considering their implication in epigenetic and post-transcriptional regulation and their cancer-/stage-specific expression profiles (Cabili et al. 2011; Iyer et al. 2015), long non-coding RNAs are very likely to play a major role in tumor progression and acquisition of resistance. They are therefore promising therapeutic-target and stage- and treatment-specific biomarkers.

Here, I summarize the current knowledge concerning the role of lncRNAs in cancer, and their potential clinical applications as biomarkers and therapeutic targets.

Regulation of lncRNA expression

As mentioned before, lncRNAs are often transcribed by the RNA polymerase II by the same mechanism producing protein-coding transcripts. However, their expression signatures have better prognostic, diagnostic, and predictive values than protein-coding signatures (Derrien et al. 2012; Brunner et al. 2012; Iyer et al. 2015; Yan et al. 2015), thus stimulating the quest for the mechanism behind such a restricted expression in specific cancer types and/or cancer stages.

Repetitive sequences, mostly belonging to the retrotransposon family, occupy 50% of the human genome and were once able to copy and paste their sequences in different areas of the genome. Having lost their capacity to transpose, these elements are now a major source of genomic innovation since they still contain cis-regulatory elements (e.g., transcriptional start sites or TSS, transcription factor binding motifs) and even genes (Göke and Ng 2016). Similarly to lncRNAs, retrotransposons display low-sequence conservation and therefore substantially differ between species (Göke and Ng 2016). Furthermore, it appears that 83.4% of long intergenic RNAs overlay at least one transposable element, while the overlap drops to 6.2% for protein-coding genes (Kelley and Rinn 2012). In particular, promoters of non-coding transcripts are highly enriched in a specific retrotransposon category known as Endogenous RetroViruses (ERVs), whose presence at the TSS seems to characterize a subclass of lncRNAs with lower and more cell-specific expression (Kelley and Rinn 2012). Endogenous retroviral sequences therefore may be a major force shaping the long non-coding RNA repertoire (Kapusta et al. 2013).

Because of their potential to induce mutations and genome instability, transposons are often silenced by a combination of sequence-specific transcription factors and epigenetic modifiers. However, hypomethylation of endogenous retroviral sequences is often observed in cancer (Szpakowski et al. 2009).

Consequently, inappropriate epigenetic activation of normally dormant, cryptic promoters, and/or enhancers derived from ERVs could be the mechanism behind the aberrant specific expression of lncRNAs in cancer (Babaian and Mager 2016).

Survival-Associated Mitochondria Melanoma-Specific Oncogenic NcRNA (SAMMSON), which is a melanoma-specific lncRNA (Leucci et al. 2016), is a potential example of an lncRNA aberrantly induced by ERV reactivation. SAMMSON expression, which is undetectable in any adult tissue, may be driven by the aberrant reactivation of the primate-specific long terminal repeat 1 A2 (LTR1A2) sequence, which sits in between the two SOX10 binding sites at the SAMMSON transcriptional start site (TSS) (Göke and Ng 2016; Babaian and Mager 2016). The functionality of LTR1A2 as a promoter seems to be supported by the fact that SAMMSON itself is conserved only in primates and could not be detected in mouse melanoma and corresponding normal skin, despite the presence of a SAMMSON homologous sequence downstream of mouse Mitf (unpublished data).

Two other examples of oncogenic lncRNAs whose expression may be driven by an ERV are urothelial cancer-associated 1 (UCA1; Wang et al. 2008) and BRAF-activated non-coding RNA (BANCR Flockhart et al. 2012). The promoter of BANCR, which is an lncRNA induced by mutated BRAF with a role in cancer cell migration (Flockhart et al. 2012), contains a MER41B LTR and actually three out of four non-coding exons are of retrotransposon origin (Kapusta et al. 2013).

Conversely, UCA1, an lncRNA originally identified as being expressed during embryonic development and in bladder cancer, consists of three exons, two or which are derived from a nested LTR element of the ERV family (Wang et al. 2008).

Lastly, human ovarian cancer-specific transcript 2 (HOST2) is a transcript not expressed in adult tissues but readily detected in ovarian cancer and not in other cancer types. The expression of this lncRNA is supposedly driven by LTR2B and the transcript is entirely made of HERV-E retroviral sequences (Babaian and Mager 2016).

Although a formal proof of the direct role of ERVs in transcription of specific lncRNAs is currently lacking, the possibility that these transcripts are the byproducts of "aberrant" genetic and/or epigenetic events reactivating retroviral elements is very attractive. If experimentally confirmed, reactivation of ERVs may provide a logical explanation for the expression of primate-restricted lncRNAs selectively in cancer tissues.

LncRNAs as lineage addiction factors

The concept that tumors arise from aberrant somatic events during normal development, and the consequent close association between cell lineage and tumorigenesis, has dominated cancer classification for over a century. As a matter of fact, the comparison of tumor-derived gene-expression patterns with those derived from corresponding normal lineage stages indicates that many tumors recapitulate normal developmental models.

It follows that lineage-survival pathways used during normal development, and aberrantly activated in cancer, profoundly influence tumor progression and metastasis and offer important therapeutic opportunities since many tumors crucially rely on those programs to survive.

An example of a key developmental pathway frequently deregulated in cancer is Notch, whose aberrant signaling is essential for t-cell acute lymphoblastic leukemia (T-ALL) maintenance. A survey of the lncRNAs expressed in T-ALL, revealed that the Notch Oncogenic Network directly regulates a consistent fraction of them. Among those, LUNAR1 was shown to be required for the efficient growth of leukemic cells since its locked nucleic acids (LNA)-mediated inhibition severely impaired cell growth *in vivo* in a xenograft model (Trimarchi et al. 2014).

A similar analysis during normal hematopoietic differentiation, in a murine model of acute myelogenous leukaemia (AML), revealed that a large number of transcripts are shared between normal developmental stages and leukemia. Among them, Pilna was further selected in a targeted shRNA screen as being essential for AML growth and survival and for normal myelogenesis, but dispensable in at least another unrelated cancer model (Delás et al. 2017).

Pilna and LUNAR1 expression and functions fall into the frame of lineage addiction, a phenomenon that can be exemplified by the role of the microphtalmia-associated transcription factor (MITF) in melanoma (Garraway and Sellers 2006). MITF is essential for the differentiation and survival of melanocytes; however, due to the genomic amplification of the 3p13–3p14 region, it is over-expressed in about 15% of melanoma patients, where it acts as an oncogene (Garraway et al. 2005).

Considering their general tissue- and stage-restricted expression and functions, lncRNAs also act as lineage-addiction factors. SAMMSON for example, is a recently characterized lncRNA transcribed downstream of the MITF gene and invariably co-amplified with it in melanoma patients (Leucci et al. 2016). Melanoma cells are addicted to SAMMSON; therefore, its depletion using antisense oligonucleotides (ASO) results in massive cell death independently of the driver mutation (Leucci et al. 2016).

Another clear lineage addiction event is the requirement of androgen receptor signaling for the development and survival of prostate epithelial cells and for prostate cancer. PCGEM1, an lncRNA expressed in the adult uniquely in the prostate and over-expressed in prostate cancer, is able to boost signaling through the androgen receptor in a hormone-dependent and -independent manner; its down-regulation strongly impairs tumor xenograft growth (Yang et al. 2013).

Another possible example of a lineage addiction lncRNA is colorectal-cancer-associated transcript 1-long isoform (CCAT1-L), which is expressed from the 8q24 region containing multiple enhancers that make tissue-specific looping with the MYC promoter (Xiang et al. 2014). CCAT1-L is specifically transcribed in colorectal cancer where it regulates transcription of MYC by chromatin looping (Xiang et al. 2014).

As exemplified by the few examples above, lineage addiction is an important source of therapeutic targets and since many lncRNAs are often specifically expressed in one particular cancer type and missing in healthy adult tissues, their systematic identification and analysis may unravel new addiction events and have important therapeutic implications.

LncRNAs as major regulators of cancer cell plasticity and drug resistance

The emergence of therapy resistance is not a surprise considering that cancer cells are constantly under evolutionary pressure dictated by the drug, the interaction with the microenvironment, and interaction between cells with different genetic and phenotypic make-up within the tumor (Baslan and Hicks 2017). Evolutionary pressure underlines the appearance of spatial and temporal intra-tumor heterogeneity that is the sum of genetic differences, cell plasticity (or the capacity of cells to shift from a transcriptional state to another one), and post-transcriptional events. All these mechanisms exert a major contribution to therapy resistance, by fuelling intra-tumor heterogeneity.

While the consequences of pre-existing and new genetic events on therapy resistance were extensively studied, the contributions of epigenetic and post-transcriptional events were recognized only recently. Accumulating evidence supports the hypothesis that, under selective pressure of the drug, cancer cells with identical genetic make-up may adopt distinct, reversible drug-tolerant states. Since the appearance of these phenotypically different states often precedes the acquisition of a stable genetic resistance (Sharma et al. 2010; Hammerlindl and Schaider 2017), a deeper understanding of drug-tolerant states and of the "non-genetic" mechanisms driving resistance is essential for the development of effective therapies. Due to their contribution to epigenetic and post-transcriptional processes, lncRNAs are very likely to participate in this process—as a matter of fact their implication in drug resistance has already been postulated (Joung et al. 2017). Many studies have suggested that deregulation of lncRNA expression could be exploited to sensitize cancer cells to targeted therapy and chemotherapy (Table 1). Already back in 2013 a very elegant study from the Rosenfeld laboratory identified and characterized two lncRNAs, called PCGEM1 and PCAT8, highly expressed in aggressive prostate cancer. PCGEM1 and PCAT8 are responsible for hormone-independent activation of the androgen receptor in castration-resistant prostate cancer and their inhibition impairs cancer growth in a xenograft model (Yang et al. 2013).

Table 1. Long non-coding RNA implicated in drug resistance.

LncRNA	Cancer	Model	Drug resistance	Target	Reference
LncARSR	Renal Cancer	Patients and Xenografts	Sunitinib	miR-34 and miR-449	Qu et al. 2016
NEAT1 (long form)	Breast, Ovarian	MCF7 2D cultures	KD Sensitizes to DNA-damaging agents	?	Adriaens et al. 2016
LINK-A	Triple negative breast cancer	Spheroids	Confers resistance AKT inhibitors	PIP_3	Lin et al. 2017
EGFR-AS1	Squamous cell carcinoma	PDX and patients	Confers resistance to Gefitinib	EGFR mRNA	Tan et al. 2017
PCAT8 and *PCGEM1*	Prostate cancer	Xenografts	Drive castration resistance	Androgen receptor	Yang et al. 2013

Long intergenic NcRNA for kinase activation (LINK-A), instead, was identified as being associated with AKT and phosphatidylinositol-3, 4,5-triphosphate (PIP_3) in triple negative breast cancer (Lin et al. 2017). Further characterization of the locus showed that it is amplified in several cancers and required to enhance recruitment of AKT to PIP3 to favor AKT activation. Inhibition of LINK-A expression by LNA-modified ASO sensitizes breast cancer spheroids to the treatment with AKT inhibitors (Lin et al. 2017).

Another example is BCAR4, which was identified in a functional screen as a transcript able to confer resistance to tamoxifen in ER+ breast cancer cells (Meijer et al. 2006). BCAR4 expression was shown to correlate with poor metastasis free survival (MFS) and progression free survival (PFS) in a cohort of 1474 estrogen receptor alpha (ERα)-positive and -negative primary breast tumors (Godinho et al., 2010).

Likewise, we recently demonstrated that the knock-down (KD) of nuclear paraspeckles assembly transcript 1 (NEAT1), a highly conserved lncRNA induced by p53, sensitized MCF7 cells to common chemotherapeutics and p53 reactivation *in vitro*, through a mechanism involving ataxia telangiectasia and rad3-related (ATR) signaling (Adriaens et al. 2016). Although very promising, these results will need to be corroborated by further studies in preclinical models and a better understanding of the NEAT1 mechanism of action.

Another strong connection between drug resistance and lncRNAs is the involvement of the last ones in the epithelial to mesenchymal transition (EMT) program and consequently in cancer stem cell biology. The EMT program (Brabletz et al., 2018) in fact plays an essential role not only during embryonic development but also in cancer stem-cell plasticity, and its partial reactivation has been connected to cancer development and progression, including therapy resistance (Shibue and Weinberg 2017). The EMT program can be triggered by several pathways including transforming-growth factor-beta (TGF-beta), wingless-related integration site (WNT), HedgeHog (HH), and hypoxia, and results in activation of three major families of transcription factors: the snail family, the zinc finger e-box-binding homeobox (ZEB) family, and the basic helix-loop-helix family (Xu et al. 2016; Shibue and Weinberg 2017). In particular, the expression of ZEB transcription factors is under tight control of the microRNA of the miR-200 family (Gregory et al. 2008). An increasing number of reports support the involvement of lncRNAs in various steps of the EMT program, thus linking lncRNA biology to drug resistance and tumor progression (Xu et al. 2016). LncRNA activated by TGF beta (LncRNA-ATB), for instance, is induced by TGF-ß in hepatocellular carcinoma where it induces EMT through up-regulation of Zeb1 and Zeb2 expression, by sponging miR-200 family members. In this case EMT was shown to contribute to invasion and metastatic colonization (Yuan et al. 2014). Another intriguing example is Zeb2 natural antisense (Zeb2NAT), a natural antisense transcript of Zeb2. By base paring to Zeb2, this lncRNA is responsible for inclusion of a large intron, containing an internal ribosome entry site (IRES), in the 5′-UTR of Zeb2 mRNA. During the induction of EMT, Zeb2NAT gets up-regulated, thus resulting in increased Zeb2 translation through the IRES and down-regulation of E-caderin (Beltran et al. 2008). Considering the implication of lncRNAs in therapy resistance and their involvement in the EMT program, there is no doubt these non-coding species would play a pivotal role in the management of therapy resistance and tumor progression.

LncRNAs as biomarkers

Early diagnosis and appropriate treatment are essential allies for the successful eradication of many types of cancers. The acquisition of resistance to targeted therapy and the emergence of new therapeutic options (e.g., immunotherapy) call for the search for new biomarkers able to stratify patients according to their responses. From this perspective, lncRNAs have great potential as prognostic, diagnostic, and predictive markers (Derrien et al. 2012; Brunner et al. 2012; Iyer et al. 2015; Yan et al. 2015), not only by virtue of their tumor-restricted expression, but also because of their stability in biological fluids, which allow the design of non-invasive diagnostic tools (table 2).

Table 2. LncRNAs as biomarkers.

LncRNA	**Cancer**	**Type of biomarker**	**Reference**
PCA3	Prostate cancer	Diagnostic marker	de Kok et al. 2002
NEAT1	Ovarian cancer	Predictive of responses to platinum	Adriaens et al. 2016
LncARSR	Renal cell carcinoma	Predictive of resistance to Sunitinib	Qu et al. 2016
EGFR-AS1	Squamous cell carcinoma	Predictive of resistance to EGFR tyrosine kinase inhibitors	Tan et al. 2017
PCAT8 and PCGEM1	Prostate cancer	Predictive of castration resistance	Yang et al. 2013
CCAT1-L	Colorectal cancer	Diagnostic marker	Nissan et al. 2012
SChLAP1	Prostate cancer	Predictive of metastasis within 10 years	Presner et al. 2014
HOTAIR	Ovarian cancer	Predictive of good response to cisplatin and bad response to carboplatin	Teschendorff et al. 2015

The stability of long transcripts in liquid biopsies is not fully understood; however, some lncRNAs travel into biological fluids protected by exosomes, thus accounting for this phenomenon. LncRNA activated in renal cell carcinoma with Sunitinib resistance (LncARSR), for instance, is packed into exosomes and can therefore be transferred to other cancer and stromal cells. The elevated expression of this transcript in therapy naive renal cell carcinoma patients correlates with poor Sunitinib responses and its inhibition by LNA-modified ASOs, and sensitizes to the treatment (Qu et al. 2016).

The use of lncRNAs as biomarkers in liquid biopsies is not new, and, as a matter of fact, prostate cancer antigen 3 (PCA3) (Bussemakers et al. 1999; de Kok et al. 2002) was approved by the Food and Drug Administration (FDA) for the diagnosis of prostate cancer already in 2012.

A promising diagnostic marker is CCAT1-L, which is exclusively transcribed in colorectal cancer. CCAT1-L is already detectable in early lesions and its levels stably increase during tumor progression (Nissan et al.

2012). Considering that CCAT1-L can be detected in the peripheral blood of 40% of patients and not in healthy controls, it could be used as a non-invasive diagnostic tool (Nissan et al. 2012). The expression of lncRNAs has not only diagnostic but also prognostic and predictive value. The expression of HOTAIR in ovarian cancer for instance, predicts poor survival for patients treated with carboplatin but good responses to cisplatin (Teschendorff et al. 2015). Two additional examples include SChLAP1, whose high expression is predictive of metastasis within 10 years in prostate cancer (Presner et al. 2014) and NEAT1 (long transcript) that reliably predicts the response to platinum-based chemotherapy in ovarian cancer (Adriaens et al. 2016).

Two approaches can be used to detect lncRNAs (figure 2): RNA fluorescence *in situ* hybridization (FISH) and/or reverse transcription (RT)-quantitative PCR/digital PCR-based assays. Both are very sensitive and can be performed using a small amount of starting material extracted from cancer tissues and/or liquid biopsies.

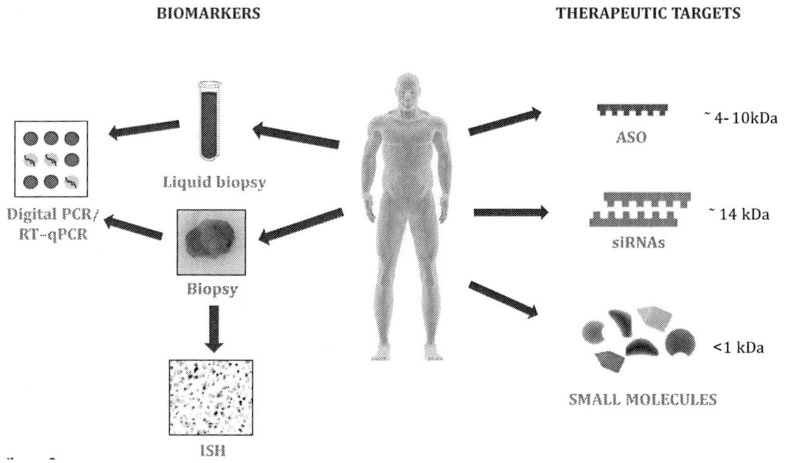

Figure 2. LncRNAs as biomarkers and therapeutic targets

LncRNA show lineage- and disease-specific expression; therefore, they have great potential as diagnostic markers and therapeutic targets. They can be detected by in situ hybridization (ISH) in regular biopsies or by digital PCR in regular and liquid biopsies. As therapeutic targets, they can be targeted by short interfering RNAs (siRNAs), antisense nucleotides (ASO), or small molecules.

RNA-ISH has two main advantages: (i) the possibility of preserving the architecture of the tissue; (ii) the option of multiplexing and combining the assay with immunohistochemistry. This technique is definitely the winning option in the case of transcripts that are unstable in body fluids and when material is scarce and does not allow subsampling. Moreover, this method would be very informative when the altered localization and not the expression of the lncRNA is indicative of the disease (or disease progression). However due to the extremely low expression of many lncRNAs and despite the several methods available to amplify the signal (e.g., tyraminde signal amplification [TSA] or RNAscope), this technique is not always an option. In such cases, the more sensitive PCR-based methods have to be selected.

Considering that lncRNA profiles can classify tumors in clinically relevant groups better than mRNAs can (Brunner et al. 2012; Iyer et al. 2015; Yan et al. 2015), that they can be detected with non-invasive tests, and that they can even be predictive of responses to therapy (Teschendorff et al. 2015; Presner et al. 2014; Adriaens et al. 2016), we can conclude that lncRNA-based toolboxes will soon become essential for pathologists and oncologists.

LncRNAs as therapeutic targets

LncRNAs are not only good biomarkers but also attractive targets for therapy (Fig. 2) and a growing number of them can impair tumor growth *in vivo*.

Among others, we showed that the systemic inhibition of SAMMSON by LNA-modified ASO in combination with Dabrafenib (an inhibitor specific to BRAFV600E retrieved in 50% of melanomas) was able to induce tumor regression of melanoma PDX models in preclinical settings (Leucci et al. 2016). Due to its melanoma-selective expression, SAMMSON therapy has the potential to avoid the emergence of toxic effects due to targeting of normal cells. Another example is THOR, a transcript transcribed from ultra conserved regions and expressed only in cancer. In a recent report, Hosono and colleagues described that the ablation of THOR in a zebrafish melanoma NRAS-driven model halts tumor formation (Hosono et al. 2017). Although preclinical evidence is still lacking, and a combination with first-line treatment agents needs to be tested, this study is very promising and suggests the possibility of targeting different cancers by inhibiting the same lncRNA.

LncRNAs can also be used to sensitize cells to several therapies and many examples have been cited in paragraph 4. One further example is epithelial growth factor receptor-anitsense 1 (EGFR-AS1), which drives resistance to EGFR in squamous cell carcinoma. Although the molecular mechanism remains to be elucidated, EGFR-AS1 expression is linked to a G/G or A/G phenotype at position 2361 in the coding region of EGFR and a combination of gefitinib with the LNA-based KD of the non-coding transcript causes tumor regression in PDX models (Tan et al. 2017).

The targeting of oncogenic lncRNAs can be achieved by silencing their expression or by pharmacological inhibition of their functions. A possible option is the use of aptamers, or nucleic acid molecules folded into a particular conformation to recognize a specific 3D-structure in a protein or an RNA (Mayer et al, 2007) conjugated to siRNAs. Aptamers can be designed to bind to predetermined surface receptors to cause internalization selectively by cancer cells/cell subpopulations (Sun et al. 2014). Despite recent progress in the field, however, siRNAs still suffer from numerous hybridization-dependent and hybridization-independent off-target effects, the last ones being caused by competing with miRNA for RISC complexes. Moreover, they are unlikely to work in certain subcellular compartments, such as mitochondria (Zhang et al. 2014).

In contrast with siRNAs, which are double-stranded, ASOs are instead single-stranded antisense molecules of 12–25 nucleotides in length; depending on the pattern of chemical modifications they carry, they can be used to cause degradation of lncRNAs or block their functions. ASOs modified only on the 5′- and 3′- ends are used for degradation of the target, while fully (LNA, 2′-OMe, constrained Ethyl, or cET . . .) modified oligos and phosphorodiamidate morpholino oligomer PMOs (and derivates) are used as steric blockers, to pharmacologically inhibit lncRNA functions. ASOs are clinically compatible, and currently four of them are FDA approved and more than 50 are in different phases of clinical trials.

Another attractive possibility to interfere with specific lncRNA functions is the use of small molecules that target RNA-protein interactions. Compared with ASOs, small molecules are of a low molecular weight (less than 1kDa) and are not charged. As a consequence, many of them can cross the blood–brain barrier; therefore, they can be an attractive option for patients with brain metastasis. However, although promising, this field is still in its infancy and information on lncRNA structures will be essential to boost it.

Degradation of RNAs by ASOs involves the recruitment of RNase H that recognizes and cleaves the DNA/RNA hybrid formed by the oligonucleotide and its target. Using RNase H for the decay, an advantage of ASOs over siRNAs is the lack of interference with the RNAi machinery. Conversely, ASOs suffer more from hybridization-independent toxicity caused by the accumulation of the drug, aptamers binding, and pro-inflammatory effects. Both ASOs and siRNAs can be further modified on the sugar moiety or on the backbone to improve the cellular uptake, the stability, and/or the specificity. Since modifications are essential to ensure the stability and delivery of the oligonucleotides, the amount of modifications that are tolerated by a given compound is definitely a criterion for the selection of the KD method. Depending on the binding to RISC for their activity, siRNAs better tolerate modifications in the sense strand, while the antisense allows only one or two internal nucleotides to be 2′-OMe (2′-O-methyl) modified (Watts et al. 2008). ASOs instead can be heavily modified and are usually flanked by stretches of modified nucleotides to prevent degradation by exonucleases.

The first modification to be introduced was the phopshotioate (PS) backbone and only oligonucleotides carrying this modification are known as first generation ASOs. Only Fomivirsen, a 21-mer that recognizes cytomegalovirus (CMV) RNA and blocks its translation, was FDA approved and commercialized for the treatment of CMV-induced retinitis (Dias and Stein 2002).

In the second generation, 2′-OMe or 2′-MOE modifications were added to the backbone to increase stability. Mipomersen, which causes the RNAse-H dependent degradation of apolipoprotein B-100, belongs to this group of ASO; it was approved by the FDA in 2012 to treat homozygous familiar hypercholesterolemia. In 2017, a splice-switching 2′-MOE ASO, called SPINRAZA, became the only approved treatment for spinal muscular atrophy (Hoy 2017). SPINRAZA base pairs with a pre-mRNA for SMN2 thus prevent exon 7 from being spliced out and SMN2 RNA from being degraded. This will allow the production of SMN that is lacking in SMA patients (Wan and Dreyfuss 2017).

The last generation of ASO incorporates further oligonucleotide modifications that increase the specificity for the target; it includes LNA, bicyclic nucleic acids (BNAs), PNAs, PMOs (or simply morpholinos), and, recently, cET from IONIS. PNA and morpholinos are not negatively charged; therefore, compared with classical ASOs, they show stronger binding to nucleic acids, but are also more difficult to deliver to the cells since they are highly hydrophobic. To overcome this problem and improve their intracellular uptake, positively charged piperazine residues and arginin-rich peptides were added to obtain modified compounds known as PMOplus and peptide-conjugated PMO (PPMOs; Kole et al. 2012). In 2016, a 30-mer PMO antisense-causing exon 51 skipping of the dystrophin mRNA (Alter et al. 2016), called eteplirsen, became the first drug for the treatment of Duchenne muscular dystrophy approved by the FDA.

Even though they are very versatile, oligonucleotides are often negatively charged and relatively bulky to cross the plasma membrane. In addition, they cannot cross the blood–brain barrier; thus, the only way to deliver them to the brain is through intrathecal injection (Juliano 2016).

Despite the recent enthusiasm triggered by the successful treatment of muscular disorders, the use of ASOs for cancer therapy is still a challenge for the future—especially as ASOs, as they are described in this review, have a clear preference for the liver and kidneys, which are also the sites of clearance of these molecules. Therefore, considering that the goal is to administrate the drug systemically for the treatment of metastatic disease, the pharmacokinetic of these molecules is a major limitation to take into account.

The future of the field is the development of conjugates that can deliver siRNAs and ASO directly to specific endocytic receptors on the cells' surface to avoid engulfment of the oligonucleotides into lysosomes (non-productive uptake). Towards this, an N-AcetylGalactosamine (GalNAc) conjugate that binds the ASialoGlycoProtein Receptor (ASGPR) has been successfully developed to specifically target hepatocytes (Prakash et al. 2014). Yet, even considering the delivery issues, we can predict that RNA-based therapeutics will be the drugs of the future, for their capacity to co-evolve easily with their target, the possibility to target lncRNA expression or function, and their relatively low costs.

Concluding remarks and future perspectives

The potential of lncRNAs in oncology as biomarkers and therapeutic targets is clearly proved by their implication in the mechanism of cancer formation and progression, their highly specific expression, and the possibility to easily design specific inhibitors. However, despite the strong evidence, we are still far away from the approval and commercialization of the first lncRNA-targeting cancer drug. To go one step further, the field needs to overcome the skepticism around systemic ASO delivery and toxicity. New chemical modifications aiming at ameliorating the bio distribution and/or the uptake of the ASOs may be needed. Moreover, from the perspective of personalized medicine, a deep understanding of specific lncRNA mechanism of action will be essential to select the most efficient inhibition strategy.

Although more advanced when compared with lncRNA therapeutics, the biomarker field is also in its infancy. The quest for lncRNAs suitable as biomarkers will need more targeted studies, designed to identify lncRNA expression profiles in specific cell subpopulations, such as drug resistant cells and/or circulating tumor cells. The rapid proliferation of single-cell studies and the adaptation of single-cell techniques to the detection of all non-coding RNA species will be a valuable tool for this purpose.

Considering the pivotal role played by lncRNAs at all stages of cancer development and progression, these efforts are worth the expected results. We can therefore anticipate that in the near future, lncRNAs will become irreplaceable tools for pathologists and oncologists.

Acknowledgments

Figures were prepared using Somersault libraries available from www.somersault1824.com. The King Baudouin Foundation supported Eleonora Leucci's work.

Conflict of interest

The authors declare that they have no conflicts of interest with respect to research, authorship, and publication of this book chapter.

Copyright and permission statement

We confirm that the materials included in this chapter do not violate copyright laws. Where relevant, appropriate permissions have been obtained from the original copyright holder(s). All original sources have been appropriately acknowledged and/or referenced.

Funding

No funding was availed for this project.

References

Adriaens, C., Standaert, L., Barra, J., Latil, M., Verfaillie, A., Kalev, P., . . . & Leucci, E. (2016). p53 induces formation of NEAT1 lncRNA-containing paraspeckles that modulate replication stress response and chemosensitivity. *Nature medicine*, 22(8), 861.

Alter, J., Lou, F., Rabinowitz, A., Yin, H., Rosenfeld, J., Wilton, S. D., . . . & Lu, Q. L. (2006). Systemic delivery of morpholino oligonucleotide restores dystrophin expression bodywide and improves dystrophic pathology. *Nature medicine*, *12*(2), 175– 77.

Arun, G., Diermeier, S., Akerman, M., Chang, K. C., Wilkinson, J. E., Hearn, S., . . . & Brogi, E. (2016). Differentiation of mammary tumors and reduction in metastasis upon Malat1 lncRNA loss. *Genes & development*, *30*(1), 34–51.

Babaian, A., & Mager, D. L. (2016). Endogenous retroviral promoter exaptation in human cancer. *Mobile DNA*, *7*(1), 24.

Bánfai, B., Jia, H., Khatun, J., Wood, E., Risk, B., Gundling, W. E., . . . & Krajewski, K. (2012). Long noncoding RNAs are rarely translated in two human cell lines. *Genome research*, *22*(9), 1646–57.

Baslan, T., & Hicks, J. (2017). Unravelling biology and shifting paradigms in cancer with single-cell sequencing. *Nature Reviews Cancer*, *17*(9), 557.

Beltran, M., Puig, I., Peña, C., García, J. M., Álvarez, A. B., Peña, R., . . . & de Herreros, A. G. (2008). A natural antisense transcript regulates Zeb2/Sip1 gene expression during Snail1-induced epithelial–mesenchymal transition. *Genes & development*, *22*(6), 756–69.

Bittker, J. A., Weiwer, M., Wei, G., Germain, A., Brown, E., Dandapani, S., . . . & Schreiber, S. L. (2011). Discovery of inhibitors of anti-apoptotic protein A1. In Probe Reports from the NIH Molecular Libraries Program [Internet]. National Center for Biotechnology Information (US).

Brabletz, T., Kalluri, R., Nieto, M. A., & Weinberg, R. A. (2018). EMT in cancer. Nature reviews cancer, *18*(2), 128.

Brunner, A. L., Beck, A. H., Edris, B., Sweeney, R. T., Zhu, S. X., Li, R., . . . & Foley, J. W. (2012). Transcriptional profiling of long non-coding RNAs and novel transcribed regions across a diverse panel of archived human cancers. *Genome biology*, *13*(8), R75.

Bussemakers, M. J., Van Bokhoven, A., Verhaegh, G. W., Smit, F. P., Karthaus, H. F., Schalken, J. A., . . . & Isaacs, W. B. (1999). DD3: A new prostate-specific gene, highly overexpressed in prostate cancer. *Cancer research*, *59*(23), 5975–79.

Byrne, A. T., Alférez, D. G., Amant, F., Annibali, D., Arribas, J., Biankin, A. V., . . . & Clarke, R. B. (2017). Interrogating open issues in cancer precision medicine with patient-derived xenografts. *Nature reviews cancer*, 17(4), 254.

Cabili, M. N., Trapnell, C., Goff, L., Koziol, M., Tazon-Vega, B., Regev, A., & Rinn, J. L. (2011). Integrative annotation of human large intergenic noncoding RNAs reveals global properties and specific subclasses. *Genes & development*, *25*(18), 1915–27.

Carrieri, C., Cimatti, L., Biagioli, M., Beugnet, A., Zucchelli, S., Fedele, S., . . . & Forrest, A. R. (2012). Long non-coding antisense RNA controls

Uchl1 translation through an embedded SINEB2 repeat. *Nature*, *491*(7424), 454–57.

Cheetham, S. W., Gruhl, F., Mattick, J. S., & Dinger, M. E. (2013). Long noncoding RNAs and the genetics of cancer. *British journal of cancer*, *108*(12), 2419–25.

Clemson, C. M., Hutchinson, J. N., Sara, S. A., Ensminger, A. W., Fox, A. H., Chess, A., & Lawrence, J. B. (2009). An architectural role for a nuclear noncoding RNA: NEAT1 RNA is essential for the structure of paraspeckles. *Molecular cell*, 33(6), 717–26.

Carninci, P., Kasukawa, T., Katayama, S., Gough, J., Frith, M. C., Maeda, N., . . . & Kodzius, R. (2005). The transcriptional landscape of the mammalian genome. *Science*, *309*(5740), 1559–63.

Delas, M. J., Sabin, L. R., Dolzhenko, E., Knott, S. R., Maravilla, E. M., Jackson, B. T., . . . & Erard, N. (2017). lncRNA requirements for mouse acute myeloid leukemia and normal differentiation. *Elife*, 6, e25607.

Derrien, T., Johnson, R., Bussotti, G., Tanzer, A., Djebali, S., Tilgner, H., . . . & Lagarde, J. (2012). The GENCODE v7 catalog of human long noncoding RNAs: analysis of their gene structure, evolution, and expression. *Genome research*, *22*(9), 1775–89.

Dias, N., & Stein, C. A. (2002). Antisense oligonucleotides: basic concepts and mechanisms. *Molecular cancer therapeutics*, *1*(5), 347–55.

Dowdy, S. F. (2017). Overcoming cellular barriers for RNA therapeutics. *Nature biotechnology*, *35*(3), 222.

Fatica, A., & Bozzoni, I. (2014). Long non-coding RNAs: new players in cell differentiation and development. *Nature reviews genetics*, *15*(1), 7–21.

Flockhart, R. J., Webster, D. E., Qu, K., Mascarenhas, N., Kovalski, J., Kretz, M., & Khavari, P. A. (2012). BRAFV600E remodels the melanocyte transcriptome and induces BANCR to regulate melanoma cell migration. *Genome research*, *22*(6), 1006–14.

Garraway, L. A., Widlund, H. R., Rubin, M. A., Getz, G., Berger, A. J., Ramaswamy, S., . . . & Lee, C. (2005). Integrative genomic analyses identify MITF as a lineage survival oncogene amplified in malignant melanoma. *Nature*, *436*(7047), 117–22.

Romanel, A., Tandefelt, D. G., Conteduca, V., Jayaram, A., Casiraghi, N., Wetterskog, D., . . . & Bianchini, D. (2015). Plasma AR and abiraterone-resistant prostate cancer. *Science translational medicine*, 7(312), 312re10–312re10.

Godinho, M. F., Sieuwerts, A. M., Look, M. P., Meijer, D., Foekens, J. A., Dorssers, L. C. J., & van Agthoven, T. (2010). Relevance of BCAR4 in tamoxifen resistance and tumour aggressiveness of human breast cancer.

British journal of cancer, 103(8), 1284–91.
Göke, J., & Ng, H. H. (2016). CTRL+ INSERT: retrotransposons and their contribution to regulation and innovation of the transcriptome. *EMBO reports*, *17*(8), 1131–44.
Gregory, P. A., Bert, A. G., Paterson, E. L., Barry, S. C., Tsykin, A., Farshid, G., . . . & Goodall, G. J. (2008). The miR-200 family and miR-205 regulate epithelial to mesenchymal transition by targeting ZEB1 and SIP1. *Nature cell biology*, 10(5), 593–601.
Guttman, M., Russell, P., Ingolia, N. T., Weissman, J. S., & Lander, E. S. (2013). Ribosome profiling provides evidence that large noncoding RNAs do not encode proteins. *Cell*, *154*(1), 240–51.
Hammerlindl, H., & Schaider, H. (2018). Tumor cell-intrinsic phenotypic plasticity facilitates adaptive cellular reprogramming driving acquired drug resistance. *Journal of cell communication and signaling*, *12*(1), 133–41.
Hezroni, H., Koppstein, D., Schwartz, M. G., Avrutin, A., Bartel, D. P., & Ulitsky, I. (2015). Principles of long noncoding RNA evolution derived from direct comparison of transcriptomes in 17 species. *Cell reports*, 11(7), 1110–22.
Holohan, C., Van Schaeybroeck, S., Longley, D. B., & Johnston, P. G. (2013). Cancer drug resistance: an evolving paradigm. *Nature reviews cancer*, 13(10), 714–26.
Hosono, Y., Niknafs, Y. S., Prensner, J. R., Iyer, M. K., Dhanasekaran, S. M., Mehra, R., . . . & Chu, S. C. (2017). Oncogenic role of THOR, a conserved cancer/testis long non-coding RNA. *Cell*, 171(7), 1559–72.
Hoy, S. M. (2017). Nusinersen: first global approval. *Drugs*, *77*(4), 473–79.
Huarte, M. (2015). The emerging role of lncRNAs in cancer. *Nature medicine*, *21*(11), 1253.
Iyer, M. K., Niknafs, Y. S., Malik, R., Singhal, U., Sahu, A., Hosono, Y., . . . & Poliakov, A. (2015). The landscape of long noncoding RNAs in the human transcriptome. *Nature genetics*, *47*(3), 199.
Joung, J., Engreitz, J. M., Konermann, S., Abudayyeh, O. O., Verdine, V. K., Aguet, F., . . . & Tseng, Y. Y. (2017). Genome-scale activation screen identifies a lncRNA locus regulating a gene neighbourhood. *Nature*, *548*(7667), 343–46.
Juliano, R. L. (2016). The delivery of therapeutic oligonucleotides. *Nucleic acids research*, *44*(14), 6518–48.
Kapusta, A., Kronenberg, Z., Lynch, V. J., Zhuo, X., Ramsay, L., Bourque, G., . . . & Feschotte, C. (2013). Transposable elements are major contributors to the origin, diversification, and regulation of vertebrate long noncoding RNAs. *PLoS genetics*, *9*(4).

Kelley, D., & Rinn, J. (2012). Transposable elements reveal a stem cell-specific class of long noncoding RNAs. *Genome biology*, *13*(11), R107.

Kino, T., Hurt, D. E., Ichijo, T., Nader, N., & Chrousos, G. P. (2010). Noncoding RNA gas5 is a growth arrest–and starvation-associated repressor of the glucocorticoid receptor. *Science signaling*, *3*(107), ra8–ra8.

De Kok, J. B., Verhaegh, G. W., Roelofs, R. W., Hessels, D., Kiemeney, L. A., Aalders, T. W., . . . & Schalken, J. A. (2002). DD3PCA3, a very sensitive and specific marker to detect prostate tumors. *Cancer research*, *62*(9), 2695–98.

Kole, R., Krainer, A. R., & Altman, S. (2012). RNA therapeutics: beyond RNA interference and antisense oligonucleotides. *Nature reviews drug discovery*, *11*(2), 125–40.

Leucci, E., Vendramin, R., Spinazzi, M., Laurette, P., Fiers, M., Wouters, J., . . . & Rogiers, A. (2016). Melanoma addiction to the long non-coding RNA SAMMSON. *Nature*, *531*(7595), 518–22.

Liu, Z., Sun, M., Lu, K., Liu, J., Zhang, M., Wu, W., . . . & Wang, R. (2013). The long noncoding RNA HOTAIR contributes to cisplatin resistance of human lung adenocarcinoma cells via downregualtion of p21WAF1/CIP1 expression. *PloS one*, *8*(10).

Lorenzen, J. M., & Thum, T. (2016). Long noncoding RNAs in kidney and cardiovascular diseases. *Nature Reviews Nephrology*, *12*(6), 360.

Mayer, G., Raddatz, M. S. L., Grunwald, J. D., & Famulok, M. (2007). RNA ligands that distinguish metabolite-induced conformations in the TPP riboswitch. *Angewandte chemie international edition*, *46*(4), 557–60.

Meijer, D., van Agthoven, T., Bosma, P. T., Nooter, K., & Dorssers, L. C. (2006). Functional screen for genes responsible for tamoxifen resistance in human breast cancer cells. *Molecular cancer research*, *4*(6), 379–86.

Mello, S. S., Sinow, C., Raj, N., Mazur, P. K., Bieging-Rolett, K., Broz, D. K., . . . & Hirose, T. (2017). Neat1 is a p53-inducible lincRNA essential for transformation suppression. *Genes & development*, *31*(11), 1095–08.

Melton, C., Reuter, J. A., Spacek, D. V., & Snyder, M. (2015). Recurrent somatic mutations in regulatory regions of human cancer genomes. *Nature genetics*, *47*(7), 710.

Nissan, A., Stojadinovic, A., Mitrani-Rosenbaum, S., Halle, D., Grinbaum, R., Roistacher, M., . . . & Bostanci, E. B. (2012). Colon cancer associated transcript-1: A novel RNA expressed in malignant and pre-malignant human tissues. *International journal of cancer*, *130*(7), 1598–606.

Ørom, U. A., & Shiekhattar, R. (2013). Long noncoding RNAs usher in a new era in the biology of enhancers. *Cell*, *154*(6), 1190–93.

Bergen, B., & Chan, T. T. (2005). Writing direction influences spatial cognition. In *Proceedings of the annual meeting of the Cognitive Science Society*, *27*(27).

Prakash, T. P., Graham, M. J., Yu, J., Carty, R., Low, A., Chappell, A., . . . & Riney, S. (2014). Targeted delivery of antisense oligonucleotides to hepatocytes using triantennary N-acetyl galactosamine improves potency 10-fold in mice. *Nucleic acids research*, *42*(13), 8796–807.

Prensner, J. R., Zhao, S., Erho, N., Schipper, M., Iyer, M. K., Dhanasekaran, S. M., . . . & Davicioni, E. (2014). RNA biomarkers associated with metastatic progression in prostate cancer: a multi-institutional high-throughput analysis of SChLAP1. *The lancet oncology*, *15*(13), 1469–80.

Quinn, J. J., & Chang, H. Y. (2016). Unique features of long non-coding RNA biogenesis and function. *Nature Reviews Genetics*, *17*(1), 47.

Sharma, S. V., Lee, D. Y., Li, B., Quinlan, M. P., Takahashi, F., Maheswaran, S., . . . & Wong, K. K. (2010). A chromatin-mediated reversible drug-tolerant state in cancer cell subpopulations. *Cell*, *141*(1), 69–80.

Shibue, T., & Weinberg, R. A. (2017). EMT, CSCs, and drug resistance: the mechanistic link and clinical implications. *Nature reviews clinical oncology*, *14*(10), 611.

Stephenson, M. L., & Zamecnik, P. C. (1978). Inhibition of Rous sarcoma viral RNA translation by a specific oligodeoxyribonucleotide. *Proceedings of the National Academy of Sciences*, *75*(1), 285–88.

Sun, H., Zhu, X., Lu, P. Y., Rosato, R. R., Tan, W., & Zu, Y. (2014). Oligonucleotide aptamers: new tools for targeted cancer therapy. *Molecular therapy: nucleic acids*, *3*, e182.

Szpakowski, S., Sun, X., Lage, J. M., Dyer, A., Rubinstein, J., Kowalski, D., . . . & Lizardi, P. M. (2009). Loss of epigenetic silencing in tumors preferentially affects primate-specific retroelements. *Gene*, *448*(2), 151–67.

Teschendorff, A. E., Lee, S. H., Jones, A., Fiegl, H., Kalwa, M., Wagner, W., . . . & Horlings, H. M. (2015). HOTAIR and its surrogate DNA methylation signature indicate carboplatin resistance in ovarian cancer. *Genome medicine*, *7*(1), 108.

Trimarchi, T., Bilal, E., Ntziachristos, P., Fabbri, G., Dalla-Favera, R., Tsirigos, A., & Aifantis, I. (2014). Genome-wide mapping and characterization of Notch-regulated long noncoding RNAs in acute leukemia. *Cell*, *158*(3), 593–606.

Ulitsky, I., Shkumatava, A., Jan, C. H., Sive, H., & Bartel, D. P. (2011). Conserved function of lincRNAs in vertebrate embryonic development despite rapid sequence evolution. *Cell*, *147*(7), 1537–50.

Wan, L., & Dreyfuss, G. (2017). Splicing-correcting therapy for SMA. *Cell*, *170*(1), 5.

Wang, F., Li, X., Xie, X., Zhao, L., & Chen, W. (2008). UCA1, a non-protein-coding RNA up-regulated in bladder carcinoma and embryo, influencing cell growth and promoting invasion. *FEBS letters*, *582*(13), 1919–27.

Watts, J. K., Deleavey, G. F., & Damha, M. J. (2008). Chemically modified siRNA: tools and applications. *Drug discovery today*, *13*(19–20), 842–855.

Xiang, J. F., Yin, Q. F., Chen, T., Zhang, Y., Zhang, X. O., Wu, Z., . . . & Yang, L. (2014). Human colorectal cancer-specific CCAT1-L lncRNA regulates long-range chromatin interactions at the MYC locus. *Cell research*, *24*(5), 513–31.

Xu, Q., Deng, F., Qin, Y., Zhao, Z., Wu, Z., Xing, Z., . . . & Wang, Q. J. (2016). Long non-coding RNA regulation of epithelial–mesenchymal transition in cancer metastasis. *Cell death & disease*, *7*(6), e2254–e2254.

Yan, X., Hu, Z., Feng, Y., Hu, X., Yuan, J., Zhao, S. D., ... & Fan, L. (2015). Comprehensive genomic characterization of long non-coding RNAs across human cancers. *Cancer cell*, *28*(4), 529–40.

Yang, L., Lin, C., Jin, C., Yang, J. C., Tanasa, B., Li, W., . . . & Evans, C. P. (2013). lncRNA-dependent mechanisms of androgen-receptor-regulated gene activation programs. *Nature*, *500*(7464), 598.

Yap, K. L., Li, S., Muñoz-Cabello, A. M., Raguz, S., Zeng, L., Mujtaba, S., . . . & Zhou, M. M. (2010). Molecular interplay of the noncoding RNA ANRIL and methylated histone H3 lysine 27 by polycomb CBX7 in transcriptional silencing of INK4a. *Molecular cell*, *38*(5), 662–74.

Yildirim, E., Kirby, J. E., Brown, D. E., Mercier, F. E., Sadreyev, R. I., Scadden, D. T., & Lee, J. T. (2013). Xist RNA is a potent suppressor of hematologic cancer in mice. *Cell*, *152*(4), 727–42.

Yuan, J. H., Yang, F., Wang, F., Ma, J. Z., Guo, Y. J., Tao, Q. F., . . . & Wang, S. B. (2014). A long noncoding RNA activated by TGF-β promotes the invasion-metastasis cascade in hepatocellular carcinoma. *Cancer cell*, *25*(5), 666–81.

Zhang, X., Zuo, X., Yang, B., Li, Z., Xue, Y., Zhou, Y., . . . & Zhang, H. (2014). MicroRNA directly enhances mitochondrial translation during muscle differentiation. *Cell*, *158*(3), 607–19.

CHAPTER 3

LONG NON-CODING RNAS IN MULTIPLE SCLEROSIS: AN UPDATE

INSHA ZAHOOR,*[1,2,3] SHAILENDRA GIRI,[1]
MIR KHURSHID IQBAL,[4]
ASHAQ HUSSAIN MIR,[3] AMRINA SHAFI,[5]
QAMAR TABAN,[3,6] AND EHTISHAMUL HAQ[2,3]

[1] Department of Neurology, Henry Ford Hospital, Detroit, Michigan, 48202, United States
[2] Bioinformatics Centre, School of Biological Sciences, University of Kashmir, Hazratbal, Srinagar, Jammu and Kashmir-190006, India
[3] Drug Therapeutics and Neurobiology Lab, Department of Biotechnology, School of Biological Sciences, University of Kashmir, Hazratbal, Srinagar, Jammu and Kashmir-190006, India
[4] Neuropharmacology Laboratory, Pharmacology Division, CSIR-Indian Institute of Integrative Medicine (IIIM), Sanat Nagar, Srinagar, Jammu and Kashmir-190005, India
[5] Signal Transduction Lab, Department of Biotechnology, School of Biological Sciences, University of Kashmir, Hazratbal, Srinagar, Jammu and Kashmir-190006, India
[6] Division of Biotechnology, Faculty of Veterinary Sciences and Animal Husbandry, Alusteng, Shuhama, SKUAST-Kashmir-190006, India
* Corresponding author:
Dr Insha Zahoor, Post-Doctoral Fellow, Department of Neurology, Research Division, Education & Research Building, Room 4023, Henry Ford Hospital, 2799 W Grand Blvd, Detroit, Michigan, 48202, United States
Tel: (313) 344-3766; Email: inshazahoor11@gmail.com

Abstract

Multiple sclerosis (MS) is a long-term immune-mediated progressive demyelinating disease of the central nervous system (CNS) that has become the major cause of neurological disability in young adults. It remains a challenging medical condition due to its enigmatic etiology. It has been time and again emphasized that one of the mechanisms behind its pathogenesis is aberrant gene expression, which is partly believed to be governed through diverse mechanisms by a new molecular species known as long non-coding ribonucleic acids (lncRNAs). Keeping in view the putative biological roles of lncRNAs in regulating multifarious functions executed by the immune system within the body, the aim of this chapter is to provide a recent and cohesive but comprehensive update on their involvement in MS pathogenesis that will indeed lay down the much needed foundation for exploiting their potential as biomarkers in the early diagnosis of this complex disease and in predicting disease activity and its course, as well as in serving as novel targets for the development of alternative therapeutic interventions for its complete cure. The recent view strongly suggests that the lncRNAs are differentially expressed in MS patients compared with control subjects and therefore play a much larger role in the pathological process of this convoluted inflammatory disease than was earlier believed to be the case. At the same time, their precise biological role, as well as the molecular mechanisms governing their role in MS pathogenesis, remains largely unexplored and needs to be further evaluated.

Keywords: biomarkers, gene expression, lncRNAs, multiple sclerosis, neurodegenerative diseases, pathogenesis

Introduction

Multiple sclerosis (MS) is classically thought to be a chronic, polygenic, multifactorial, heterogeneous, and auto-inflammatory disease affecting the primary central nervous system (CNS) of the body (Compston and Coles 2008). It is characterized by the loss of myelin (demyelination) from the nerve fibers due to an uncharacteristic or more appropriately dysregulated immune response against the myelin sheath and myelin-producing cells (Compston and Coles 2002; Peterson and Fujinami 2007). The diagrammatic representation of the demyelination process in MS is shown in figure 1. Depending on the location of the damage, it exhibits diverse symptomatology (Compston and Coles 2008). At the cellular level, there occurs the accumulation of disability and/or damage over time due to the inability of the inbuilt repair system to replace the lost myelin effectively through the process of remyelination and upon repeated damage; this ultimately leads to the formation of a lesion at the damaged site (Chari 2007).

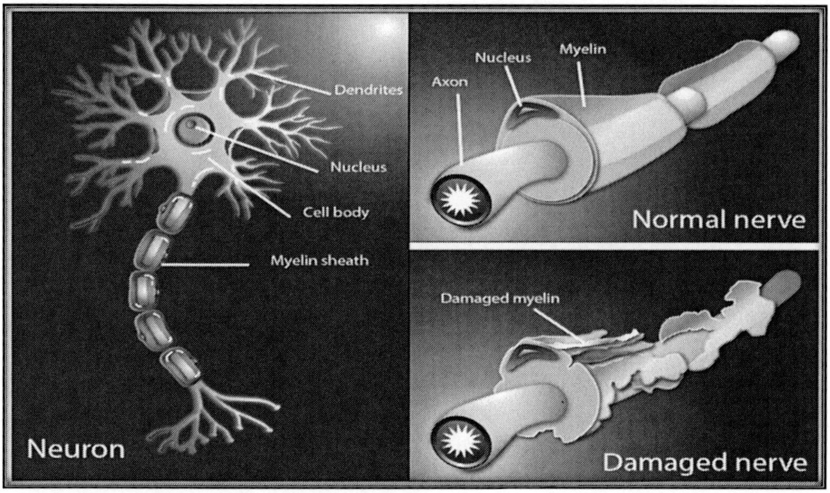

Figure 1. MS develops due to an immune attack on the myelin sheath coating nerves within the CNS. (Source: https://www.onhealth.com/content/1/multiple_sclerosis_ms.)

The hallmark of MS etiology is the amalgamation of several genetic and environmental players that interact in a complex way in triggering the disease (Oksenberg and Baranzini 2010; Sayad et al. 2017; Taheri et al., 2017) (figure 2). Examples of such factors include viral infections like Epstein Barr virus (EBV) (Ascherio and Munger 2007a; Compston and Coles 2008), smoking (Ascherio and Munger 2007b), toxic chemicals (Marrie 2004), and vitamin D deficiency (Ascherio, Munger, and Simon 2010; Ascherio and Munger 2007b; Holick et al. 2015; Zahoor and Haq 2017) (figure 2). A large body of evidence suggests the predominant role of diverse genetic elements in MS predisposition (Dyment, Ebers, and Sadovnick 2004) and, over time, variations in some major candidate genes such as human leukocyte antigen (HLA) have come to light as causes of this chronic disease (Baranzini 2011; Compston and Coles 2008). Apart from this, genome-wide association studies have linked MS susceptibility with variations in several non-HLA genes like interleukin-2 receptor alpha (IL2RA) (Weber et al. 2008), tumor necrosis factor receptor superfamily member 1A (TNFRSF1A) (De Jager et al. 2009), 25-hydroxyvitamin D-1 alpha-hydroxylase (CYP27B1) (Sundqvist et al. 2010), and interleukin-7 receptor alpha (IL7RA) (Lundmark et al. 2007); however, each confers only a small effect (figure 2).

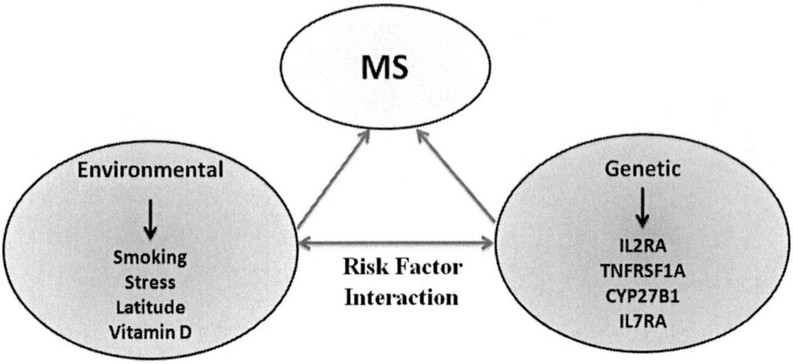

Figure 2. Interactions between different environmental and genetic risk factors lead to MS.

Clinically, on the basis of disease activity and its progression, MS has been categorized into four different courses; namely clinically isolated syndrome (CIS), relapsing-remitting (RR), secondary progressive (SP), and primary progressive (PP) (Lublin et al. 2014). Among them, the most common course is the RR type, which is found in 85% of patients across the globe. The future course of MS is highly unpredictable, making its prognosis very difficult; as a result, it has become the primary cause of disability in young adults, which has resulted in an enhanced disease burden globally (Compston and Coles 2008). Its ever-increasing prevalence and enigmatic etiology make it a key topic of discussion for clinicians, researchers, and global leaders. Over the past decade, there has been a paradigm shift in understanding the molecular mechanisms behind the disease process as research on MS has rapidly progressed, thereby leading to critical advances in understanding its pathogenesis (Eftekharian et al. 2016). Interestingly, there have been numerous reports suggesting a new class of macromolecules that play a critical role in the pathological process of a wide variety of human diseases including neurological, oncological, and cardiovascular diseases and diabetes, with tremendous potential for serving as biomarkers in predicting disease courses and activities (Lee et al. 2015; Li and Wang 2016; Liu et al. 2014; Shi and Yang 2016; Soudyab, Iranpour, and Ghafouri-Fard 2016; Wan, Su, and Zhuo 2017; P. Wu et al. 2013; Ying et al. 2013). These biomarkers include a novel class of non-coding ribonucleic acids (ncRNAs) known as long non-coding RNAs (lncRNAs), which are a group of transcripts with a length of more than 200 nucleotides that lack protein-coding ability (Ip and Nakagawa 2012). They have been very well documented, playing a regulatory role in gene expression, thereby modulating a wide continuum of physiological processes including apoptosis, proliferation, development, differentiation, and development, as well as maintenance of immune function at the epigenetic, transcriptional, and post-transcriptional levels (Kretz et al. 2013; P. Wang et al. 2014; Ying et al. 2013; Y. Zhang and Cao 2016). Consequently, they play a crucial role in maintaining cellular homeostasis due to which they have emerged as new players in the maintenance, functioning, and development of CNS (Tan et al. 2013; G. C. Wu et al. 2015).

Over the last ten years, lncRNAs have really gained tremendous importance in the development of neurodegenerative diseases including Parkinson's disease (PD), amyotrophic lateral sclerosis (ALS), Huntington's disease (HD), Alzheimer's disease (AD), frontotemporal lobar degeneration (FTLD), multiple system atrophy (MSA), and glaucoma (Chen and Zhou 2017; Riva, Ratti, and Venturin 2016; Wan et al. 2017; D. Q. Wang et al. 2018). Interestingly, the recent entrant in this category is MS (Bina et al. 2017; Eftekharian et al. 2017; Pahlevan et al. 2018; Santoro et al. 2016; F. Zhang et al. 2016). However, the exact mechanism of action for these lncRNAs is not well documented and despite escalating research across the globe, this process still remains poorly understood. This chapter focuses on the role of lncRNAs in MS pathogenesis by taking into consideration the cumulative outcome of the latest studies, which has significant potential in disease diagnosis and therapeutics. This chapter is the first of its kind to highlight the role of lncRNAs in MS development with a focus on FTLD, MSA, and glaucoma. This will lay down a foundation for the investigation of novel elements involved in the pathogenesis of neurodegenerative diseases, thereby contributing to disease development at the molecular level, which could eventually provide a clear picture of the disease process.

LncRNAs in MS

Emerging evidence has revealed that lncRNAs may play a crucial role in autoimmunity by regulating immune response at different levels of immunological cascades including development, differentiation, activation, and expression of immune cells (including T cells, B cells, macrophages, and NK cells) by influencing epigenetic processes such as the modification of the chromatin state; this is suggestive of their potential impact on the pathological process of MS (Sigdel et al. 2015; Spurlock et al. 2015; Y. Zhang and Cao 2016) (figure 3). At the same time their expression profile in autoimmunity vis-à-vis autoimmune disease is not clearly known and this needs to be explored. The putative mechanisms involved in lncRNAs being regarded as culprits in the pathophysiology of and susceptibility to develop neurodegenerative diseases include variations in lncRNA genes, epigenetic dysregulation, a link between lncRNAs and disease genes at the genomic level, and the involvement of lncRNA in disease pathological networks (Qureshi and Mehler 2012, 2013). Incidentally, multiple lines of evidences have strongly advocated the importance of these lncRNAs in MS pathogenesis, which has also resulted in active debates among researchers, following which there has been a series of research papers based on the expression profile of lncRNAs in MS (Bina et al. 2017; Eftekharian et al. 2017; Pahlevan et al. 2018; Santoro et al. 2016; F. Zhang et al. 2016). So far, all the studies on lncRNAs and MS have been carried out mostly on patients with an RR disease course.

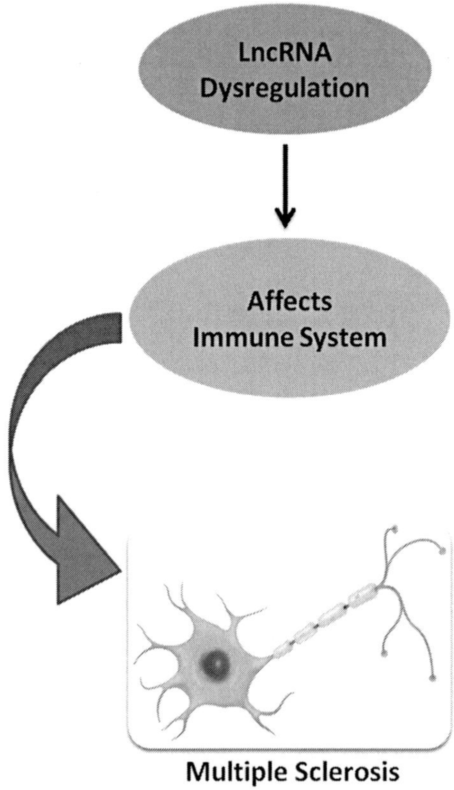

Figure 3. Dysregulation of lncRNAs causes MS by affecting regulation of various immunological functions including autoimmunity.

A comparative genome-wide study by Zhang et al. in 2015 evaluated the expression profile of lncRNAs in the peripheral blood mononuclear cells (PBMCs) of 26 MS patients and age- and gender-matched controls by means of the RNA microarray method (F. Zhang et al. 2016). Among 26 patients, only 6 were randomly selected for microarray analysis. Microarray data analyses revealed significant differential expression of 2742 lncRNAs, among which 2353 lncRNAs were up-regulated and 389 down-regulated with a more than six-fold difference in 137 of them; 37 had a difference of greater than eight-fold in the expression level. Real-time polymerase chain reaction (RT-PCR) results were relatively consistent with the microarray data, confirming the same pattern of regulation and differential expression in only six of these lncRNAs among MS cases and controls. The lncRNA functions for the target genes were predicted using *Gene Ontology* (GO) and *Kyoto Encyclopedia of Genes and Genomes* (KEGG) analyses; simultaneously, regulatory relationships between them were also analyzed using the "cis" and "trans" models. Different bioinformatics tools were used to predict the impact of the dysregulated lncRNAs on their target genes for deciphering their probable role in MS. Among aberrant lncRNAs, 10 were predicted to have 10 cis-regulated target genes, and 33 might regulate their trans target genes. It was revealed that lncRNAs ENST00000491934.2 and HSP90AA4P might be involved in the pathophysiology of MS by regulating the expression of Homeobox C9 (HOXC9) and DNA damage inducible transcript 4 (DDIT4) genes, respectively, in a cis-oriented manner; the other lncRNA, XR_132575.3, could affect the expression of NKx2-5 in the trans approach. These lncRNAs could exert their influences on target genes in conjunction with other transcription factors. The findings infer the aberrant expression of lncRNAs during the initiation as well as the propagation of the immune response (F. Zhang et al. 2016), thereby providing major evidence for the potential role of lncRNAs and their target genes in MS pathology.

In an attempt to explore the link between lncRNAs and MS, another expression profile study by Santoro et al. screened a subset of lncRNAs known to play a crucial role in inflammatory and autoimmune responses by controlling the expression of pro-inflammatory, anti-inflammatory genes, and microRNA involved in mediating such responses (Santoro et al. 2016). The expression profiles of 84 lncRNAs were assessed in the serum of 12 MS patients and compared with profiles of age- and gender-matched controls and idiopathic inflammatory myopathy (IIM) patients by means of RT-PCR. Among these screened lncRNAs, only 3 were found to be significantly up-regulated in patients compared with controls, namely nuclear paraspeckle assembly transcript 1 (NEAT1), taurine up-regulated 1 (TUG1), and 7SK small nuclear (RN7SK RNA); if we go by the literature, they have been found to be involved in neurodegenerative diseases (Nishimoto et al. 2010). Moreover, a study by Bina et al. assessed the expression level and function of lnc-interleukin 7 receptor (lnc-IL-7R) in regulating the expression of membrane-bound (IL-7RB) and soluble (IL-7RS) isoforms of IL-7R alpha in 36 MS patients and 30 controls by means of RT-PCR (Bina et al. 2017). The outcome of the study was insignificant as there was no difference in the expression levels of IL-7RB and IL-7RS and lnc-IL-7R between patients or controls. However, a significant association was observed between the expression levels of IL-7RB and lnc-IL-7R, IL-7RS and lnc-IL-7R, and IL-7RB and IL-7RS in both study groups. This has created new information regarding the mechanism of regulation being exercised by lncRNAs in controlling the gene expression (Bina et al. 2017).

In order to elucidate the role of lncRNAs in MS, a comparative study by Eftekharian et al. evaluated the expression of a subset of lncRNAs in the PBMCs of 50 Iranian MS patients and age- and gender-matched controls using quantitative RT-PCR (qRT-PCR) (Eftekharian et al. 2017). The lncRNAs screened in the study—namely, TNF-α and heterogeneous nuclear ribonucleoprotein L (THRIL), Fas cell surface death receptor-antisense 1 (FAS-AS1), and plasmacytoma variant translocation 1 (PVT1)—were believed to have a regulatory role in the functioning of the immune system. They demonstrated significant dysregulation of these 3 lncRNAs with the down-regulation of PVT1 and FAS-AS1 and the up-regulation of THRIL in patients compared with controls. PVT1 expression was found to be predominately lower in female MS patients compared with female controls. In contrast to this, FAS-AS1 expression showed no gender bias in patients as it was lower in both genders. Similarly, there was a higher expression of THRIL in all patients regardless of gender. Furthermore, no significant association was observed in terms of lncRNA expression levels and age at MS onset, where lncRNA expression levels and the degree of disability in patients was measured in the form of a score known as the expanded disability status scale (EDSS). However, a moderate association was found between lncRNA expression levels and disease duration.

Furthermore, it is well known that vitamin D plays an immunomodulatory role in MS; therefore, it is prescribed as a supplement to patients with low vitamin D levels (Holick et al. 2015; Zahoor and Haq 2017). At the same time, its metabolism is putatively regulated by diverse mechanisms including the epigenetic mode of regulation, which makes lncRNAs potent targets for study as regulators (Saccone, Asani, and

Bornman 2015). In this context, a study by Pahlevan et al. examined the role of HOX transcript antisense intergenic RNA (HOTAIR) and ANRIL (antisense lncRNA of the INK4 locus) in the pathological process of MS in line with vitamin D levels in patients with vitamin D deficiency and patients supplemented with vitamin D (Pahlevan et al. 2018). So far, this is the first expression study of HOTAIR and ANRIL lncRNAs in relation to vitamin D supplementation. Studies have shown that both these lncRNAs alter chromatin structure by influencing the chromatin-remodeling complexes through histone modifications (Kotake et al. 2011; Rinn et al. 2007; Yap et al. 2010). In the report by Pahlevan et al. it was observed that the expression levels of HOTAIR were significantly higher in the PBMCs of vitamin D–deficient MS patients before supplementation compared with the controls and there was a subsequent decrease in the levels after vitamin D supplementation. However, the exact mechanism and role of this up-regulation in the pathogenesis of MS remain elusive. It is believed that vitamin D exerts its effects on HOTAIR expression through c-Myc (Ma et al. 2014; Piek et al. 2010): in this study, it was found significantly up-regulated after supplementing patients with vitamin D. As far as gender bias in the expression levels of HOTAIR is concerned, it was over-expressed in female patients before vitamin D supplementation in comparison with males and controls. This reflects the involvement of the endocrine system and requires further evaluation. There was no significant findings for ANRIL and all the above data so far point towards the potential role of HOTAIR in MS pathogenesis, which could mediate this process in conjunction with vitamin D through different putative epigenetic mechanisms including histone modification, chromatin remodeling, or miRNA regulation (Pahlevan et al. 2018). This may eventually influence myelination within CNS and other immune-mediated processes that are hallmarks in the pathological process of MS. At the same time, the complete mechanism behind the role of vitamin D in governing HOTAIR-based processes needs further validation.

The comparison made between some of the major studies based on the expression analysis of lncRNAs in MS is outlined in table 1.

Table 1. Comparative analysis between some major expression studies of lncRNAs in MS.				
S. No.	**No. of patients screened**	**No. of LncRNAs/ ID**	**Outcome**	**Reference**
1.	26	2353 389	Up-regulation Down-regulation	(F. Zhang et al. 2016)
2.	12	NEAT1 TUG1 RN7SK RNA	Up-regulation	(Santoro et al. 2016)
3.	36	lnc-IL-7R	Insignificant	(Bina et al. 2017)
4.	50	THRIL FAS-AS1 PVT1	Up-regulation Down-regulation Down-regulation	(Eftekharian et al. 2017)
5.	32	HOTAIR ANRIL	Up-regulation Insignificant	(Pahlevan et al. 2018)

NEAT1: nuclear paraspeckle assembly transcript 1; TUG1: taurine up-regulated 1; RN7SK RNA: 7SK small nuclear; lnc-IL-7R: lnc-interleukin 7 receptor; THRIL: TNF-α and heterogeneous nuclear ribonucleoprotein L; FAS-AS1: Fas cell surface death receptor-antisense 1; PVT1: plasmacytoma variant translocation 1; HOTAIR: HOX transcript antisense intergenic RNA; ANRIL: antisense lncRNA of the INK4 locus.

LncRNAs in other diseases

LncRNAs have also been found to be involved in the pathogenesis of other neurological disorders, which include MSA, FTLD, and glaucoma (P. Wu et al. 2013). So far, there has not been much research on their role in these diseases (Wan et al. 2017). MSA is a rare progressive disease commonly known as Shy–Drager syndrome that affects both the autonomic nervous system and movement of the body. There occurs rapid degeneration of neurons in the substantia nigra, striatum, inferior olivary nucleus, and cerebellum giving rise to a wide continuum of signs and symptoms including Parkinsonian features, tremors, slow movement, spasticity, and ataxia. It shares some similarity with other neurodegenerative diseases including spinocerebellar ataxia type 8 (SCA8); in line with this (Wan et al. 2017), a report by Munhoz et al. looked for the presence of CTA/CTG expansions at the SCA 8 locus in 10 MSA patients out of which 9 patients carried combined repeats in the normal range of 20 to 28 and only one patient showed 22/76 repeats, thereby exceeding normal range; however, from the clinical point of view, there was no significant difference between patients in respect of the disease presentation (Munhoz et al. 2009). The overall findings of the study do not provide significant information on the involvement of lncRNAs in MSA development.

Another rare progressive neurodegenerative disease of the brain is FTLD in which there occurs severe degeneration and/or atrophy of the frontal and temporal lobes of the brain. Depending on the type of cytoplasmic inclusions present in the brain, 3 different histological types have been observed: FTLD-tau, FTLD-TDP, and FTLD-FUS (Mackenzie et al. 2011). FTLD-TDP and FTLD-FUS pathologies are characterized by the accumulation of the RNA-binding proteins (RBPs) TDP-43 (transactive response DNA binding protein 43 kDa) and FUS/TLS (fused in sarcoma/translocated in sarcoma), respectively. It has been recently put forward that lncRNAs possess sites for binding of FTLD-related proteins and that binding of TDP-43 and FUS/TLS affects the cellular functioning of lncRNAs by controlling their transcription and stability (Lourenco et al. 2015). Furthermore, studies have also revealed the involvement of NEAT1-2 in FTLD pathology by experimental demonstration of direct binding between NEAT1-2 and TDP-43 and FUS/TLS (Nishimoto et al. 2010). However, there is still a lack of concrete evidence behind the exact role of lncRNAs in the development of FTLD.

Another condition, glaucoma, is actually a complex group of heterogeneous ocular diseases characterized by damage to the optic nerve and vision loss (Mantravadi and Vadhar 2015). Its pathophysiology is similar to other neurodegenerative diseases like Parkinson's disease and Alzheimer's disease (Danesh-Meyer and Levin 2015). A report by a US-based research group, Pasquale et al., explored the correlation between 10 single nucleotide polymorphisms (SNPs) in ANRIL lncRNA, also known as CDKN2B-AS1, and glaucoma features in 976 primary open-angle glaucoma (POAG) patients from the Glaucoma Genes and Environment (GLAUGEN) study and 1971 patients from the National Eye Institute Glaucoma Human Genetics Collaboration (NEIGHBOR) consortium (Pasquale et al., 2013). Out of 10 SNPs, 9 were found to have a protective nature and showed an association with a reduced risk of developing glaucoma whereas 1 SNP was found to be disease-causing; it could therefore be suggested that these SNPs not only affect the disease risk but also create optic nerve degeneration in POAG patients. On similar lines, another study by Nakano et al. examined genome-wide association studies in Japanese and identified five common variants in ANRIL showing significant association with optic-nerve vulnerability in 2219 glaucoma patients from Japan, which proved to be consistent with the findings from the study conducted on Caucasians (Nakano et al., 2012). Other studies have shown the role of lncRNA LOXL1 antisense RNA 1 (LOXL1-AS1) in glaucoma as it has been reported to play an important role in response to different cellular stresses such as oxidative stress in lens epithelial cells and cyclic mechanical stress in Schlemm's canal (Aboobakar et al. 2017; Hauser et al. 2015). Similarly, there are other genetic elements that have been shown to play a role in glaucoma pathogenesis. These include metastasis-associated lung adenocarcinoma transcript 1 (MALAT1), also known as NEAT2; optineurin; ASMM10P055228; and ASMM10P040128 (Chi et al. 2010; Fuse et al. 2004; H. B. Li et al. 2017; Li, Jin et al. 2017; Rezaie et al. 2002; Sarfarazi and Rezaie 2003). Despite ongoing studies, the exact molecular mechanisms behind their association remain elusive (Cissé, Bai, and Meng 2018; Ng et al. 2014).

Conclusion and future perspectives

LncRNAs are expressed copiously in the brain, especially the brain-specific subset of lncRNA (Derrien et al. 2012). Although, only a few lncRNAs have been found to be associated with the development and functioning of the human brain, advanced molecular studies have revealed their key role in the diagnosis, pathogenesis, and prognosis of multiple neurodegenerative diseases (Lee et al. 2015; Vucicevic, Schrewe, and Orom 2014; Wan et al. 2017; P. Wu et al. 2013). With the advent in RNA-based technologies and in silico tools, studies have strongly implicated the aberrant expression of lncRNAs as an underlying cause for MS development. However, the studies aiming to uncover the exact mechanism behind their involvement in its pathogenesis are still in their infancy. There have been very few studies based on the expression profile of different lncRNAs in the serum of patients suffering from MS. At the same time, their pattern of expression, key target genes, and precise mechanism of action in the pathological process of MS remains elusive. The expression analysis of lncRNAs during different courses of MS as well as different stages of the immune response would definitely provide important clues for novel drug development approaches. It could also pave the way for evaluating treatment response in patients; thereby aiding in the development of personalized medicine for achieving an appropriate response in patients with high efficacy, which will reduce the cost as well as the adverse side effects of drugs. So far, there has been no such exclusive draft based on the importance of lncRNAs in MS development. On the basis of these grounds, and in view of the tissue-specific expression of lncRNAs, this chapter will ultimately provide a much-needed platform for new investigations that could manipulate their functions and use them as molecular markers leading to novel insights into the diagnosis, etiology, pathogenesis, prognosis, and treatment of autoimmune diseases especially MS.

Conflict of interest

The authors declare that they have no conflicts of interest with respect to research, authorship, and publication of this book chapter.

Copyright and permission statement

We confirm that the materials included in this chapter do not violate copyright laws. Where relevant, appropriate permissions have been obtained from the original copyright holder(s). All original sources have been appropriately acknowledged and/or referenced.

Funding

No funding was received for this project.

Acknowledgements

We are highly indebted to the Bioinformatics Centre, University of Kashmir, for providing their services while drafting this chapter.

References

Aboobakar, I. F., Johnson, W. M., Stamer, W. D., Hauser, M. A., & Allingham, R. R. (2017). Major review: Exfoliation syndrome; advances in disease genetics, molecular biology, and epidemiology. *Experimental eye research*, *154*, 88–103.

Ascherio, A., & Munger, K. L. (2007a). Environmental risk factors for multiple sclerosis. Part I: the role of infection. *Annals of neurology*, *61*(4), 288–99.

Ascherio, A., & Munger, K. L. (2007b). Environmental risk factors for multiple sclerosis. Part II: Noninfectious factors. *Annals of neurology*, *61*(6), 504–13.

Ascherio, A., Munger, K. L., & Simon, K. C. (2010). Vitamin D and multiple sclerosis. *Lancet neurology*, *9*(6), 599–612.

Baranzini S. E. (2011). Revealing the genetic basis of multiple sclerosis: are we there yet? *Current opinion in genetics & development*, *21*(3), 317–24.

Bina, P., Pahlevan Kakhki, M., Sahraian, M. A., & Behmanesh, M. (2017). The expression of lnc-IL-7R long non-coding RNA dramatically correlated with soluble and membrane-bound isoforms of IL-7Ra gene in multiple sclerosis patients. *Neuroscience letters*, *642*, 174–78.

Chari D. M. (2007). Remyelination in multiple sclerosis. *International review of neurobiology*, *79*, 589–620.

Chen, Y., & Zhou, J. (2017). LncRNAs: macromolecules with big roles in neurobiology and neurological diseases. *Metabolic brain disease*, *32*(2), 281–91.

Chi, Z. L., Akahori, M., Obazawa, M., Minami, M., Noda, T., Nakaya, N., … & Iwata, T. (2010). Overexpression of optineurin E50K disrupts Rab8 interaction and leads to a progressive retinal degeneration in mice. *Human molecular genetics*, *19*(13), 2606–15.

Cissé, Y., Bai, L., & Meng, T. (2018). LncRNAs in genetic basis of glaucoma. *BMJ open ophthalmology*, *3*(1), e000131.

Compston, A., & Coles, A. (2002). Multiple sclerosis. *Lancet*, *359*(9313), 1221–31.

Compston, A., & Coles, A. (2008). Multiple sclerosis. *Lancet*, *372*(9648), 1502–17.

Danesh-Meyer, H. V., & Levin, L. A. (2015). Glaucoma as a neurodegenerative disease. *Journal of neuro-ophthalmology*, *35 Suppl 1*, S22–S28.

De Jager, P. L., Jia, X., Wang, J., de Bakker, P. I., Ottoboni, L., Aggarwal, N. T., … & Oksenberg, J. R. (2009). Meta-analysis of genome scans and replication identify CD6, IRF8 and TNFRSF1A as new multiple sclerosis susceptibility loci. *Nature genetics*, *41*(7), 776–82.

Derrien, T., Johnson, R., Bussotti, G., Tanzer, A., Djebali, S., Tilgner, H., . . . & Guigó, R. (2012). The GENCODE v7 catalog of human long noncoding RNAs: analysis of their gene structure, evolution, and expression. *Genome research*, *22*(9), 1775–89.

Dyment, D. A., Ebers, G. C., & Sadovnick, A. D. (2004). Genetics of multiple sclerosis. *Lancet neurology*, *3*(2), 104–10.

Eftekharian, M. M., Ghafouri-Fard, S., Soudyab, M., Omrani, M. D., Rahimi, M., Sayad, A., . . . & Taheri, M. (2017). Expression Analysis of Long Non-coding RNAs in the Blood of Multiple Sclerosis Patients. *Journal of molecular neuroscience*, *63*(3–4), 333–41.

Eftekharian, M. M., Noroozi, R., Sayad, A., Sarrafzadeh, S., Toghi, M., Azimi, T., . . . & Mirfakhraie, R. (2016). RAR-related orphan receptor

A (RORA): A new susceptibility gene for multiple sclerosis. *Journal of the neurological sciences*, *369*, 259–62.

Fuse, N., Takahashi, K., Akiyama, H., Nakazawa, T., Seimiya, M., Kuwahara, S., & Tamai, M. (2004). Molecular genetic analysis of optineurin gene for primary open-angle and normal tension glaucoma in the Japanese population. *Journal of glaucoma*, *13*(4), 299–303.

Hauser, M. A., Aboobakar, I. F., Liu, Y., Miura, S., Whigham, B. T., Challa, P., . . . & Allingham, R. R. (2015). Genetic variants and cellular stressors associated with exfoliation syndrome modulate promoter activity of a lncRNA within the LOXL1 locus. *Human molecular genetics*, *24*(22), 6552–63.

Holick, M. F., Cook, S., Suarez, G., & Rametta, M. (2015). Vitamin D deficiency and possible role in multiple sclerosis. *European neurological review, 10*(2), 131–38.

Ip, J. Y., & Nakagawa, S. (2012). Long non-coding RNAs in nuclear bodies. *Development, growth & differentiation*, *54*(1), 44–54.

Kotake, Y., Nakagawa, T., Kitagawa, K., Suzuki, S., Liu, N., Kitagawa, M., & Xiong, Y. (2011). Long non-coding RNA ANRIL is required for the PRC2 recruitment to and silencing of p15(INK4B) tumor suppressor gene. *Oncogene*, *30*(16), 1956–62.

Kretz, M., Siprashvili, Z., Chu, C., Webster, D. E., Zehnder, A., Qu, K., . . . & Khavari, P. A. (2013). Control of somatic tissue differentiation by the long non-coding RNA TINCR. *Nature*, *493*(7431), 231–35.

Lee, D. Y., Moon, J., Lee, S. T., Jung, K. H., Park, D. K., Yoo, J. S., . . . & Lee, S. K. (2015). Dysregulation of long non-coding RNAs in mouse models of localization-related epilepsy. *Biochemical and biophysical research communications*, *462*(4), 433–40.

Li, H. B., You, Q. S., Xu, L. X., Sun, L. X., Abdul Majid, A. S., Xia, X. B., & Ji, D. (2017). Long non-coding RNA-MALAT1 mediates retinal ganglion cell apoptosis through the PI3K/Akt signaling pathway in rats with glaucoma. *Cellular physiology and biochemistry*, *43*(5), 2117–32.

Li, Y., & Wang, X. (2016). Role of long noncoding RNAs in malignant disease (Review). *Molecular medicine reports*, *13*(2), 1463–69.

Li, Y., Jin, L., Dong, A., Zhou, X., & Yuan, H. (2017). Microarray expression profile analysis of long non-coding RNAs in optineurin E50K mutant transgenic mice. *Molecular medicine reports*, *16*(2), 1255–61.

Liu, J. Y., Yao, J., Li, X. M., Song, Y. C., Wang, X. Q., Li, Y. J., . . . & Jiang, Q. (2014). Pathogenic role of lncRNA-MALAT1 in endothelial cell dysfunction in diabetes mellitus. *Cell death & disease*, *5*(10), e1506.

Lourenco, G. F., Janitz, M., Huang, Y., & Halliday, G. M. (2015). Long noncoding RNAs in TDP-43 and FUS/TLS-related frontotemporal lobar degeneration (FTLD). *Neurobiology of disease*, *82*, 445–54.

Lublin, F. D., Reingold, S. C., Cohen, J. A., Cutter, G. R., Sørensen, P. S., Thompson, A. J., . . . & Polman, C. H. (2014). Defining the clinical course of multiple sclerosis: the 2013 revisions. *Neurology*, *83*(3), 278–86.

Lundmark, F., Duvefelt, K., Iacobaeus, E., Kockum, I., Wallström, E., Khademi, M., . . . & Hillert, J. (2007). Variation in interleukin 7 receptor alpha chain (IL7R) influences risk of multiple sclerosis. *Nature genetics*, *39*(9), 1108–13.

Ma, M. Z., Li, C. X., Zhang, Y., Weng, M. Z., Zhang, M. D., Qin, Y. Y., … & Quan, Z. W. (2014). Long non-coding RNA HOTAIR, a c-Myc activated driver of malignancy, negatively regulates miRNA-130a in gallbladder cancer. *Molecular cancer*, *13*, 156.

Mackenzie, I. R., Neumann, M., Baborie, A., Sampathu, D. M., Du Plessis, D., Jaros, E., . . . & Lee, V. M. (2011). A harmonized classification system for FTLD-TDP pathology. *Acta neuropathologica*, *122*(1), 111–13.

Mantravadi, A. V., & Vadhar, N. (2015). Glaucoma. *Primary care*, *42*(3), 437–49.

Marrie R. A. (2004). Environmental risk factors in multiple sclerosis aetiology. *Lancet neurology*, *3*(12), 709–18.

Munhoz, R. P., Teive, H. A., Raskin, S., & Werneck, L. C. (2009). CTA/CTG expansions at the SCA 8 locus in multiple system atrophy. *Clinical neurology and neurosurgery*, *111*(2), 208–10.

Nakano, M., Ikeda, Y., Tokuda, Y., Fuwa, M., Omi, N., Ueno, M., . . . & Tashiro, K. (2012). Common variants in CDKN2B-AS1 associated with optic-nerve vulnerability of glaucoma identified by genome-wide association studies in Japanese. *PloS one*, *7*(3), e33389.

Ng, S. K., Casson, R. J., Burdon, K. P., & Craig, J. E. (2014). Chromosome 9p21 primary open-angle glaucoma susceptibility locus: a review. *Clinical & experimental ophthalmology*, *42*(1), 25–32.

Nishimoto, Y., Ito, D., Yagi, T., Nihei, Y., Tsunoda, Y., & Suzuki, N. (2010). Characterization of alternative isoforms and inclusion body of the TAR DNA-binding protein-43. *Journal of biological chemistry*, *285*(1), 608–19.

Oksenberg, J. R., & Baranzini, S. E. (2010). Multiple sclerosis genetics—is the glass half full, or half empty? *Nature reviews: neurology*, *6*(8), 429–37.

Pahlevan Kakhki, M., Nikravesh, A., Shirvani Farsani, Z., Sahraian, M. A., & Behmanesh, M. (2018). HOTAIR but not ANRIL long non-coding RNA contributes to the pathogenesis of multiple sclerosis. *Immunology*, *153*(4), 479–87.

Pasquale, L. R., Loomis, S. J., Kang, J. H., Yaspan, B. L., Abdrabou, W., Budenz, D. L., . . . & Wiggs, J. L. (2013). CDKN2B-AS1 genotype-glaucoma feature correlations in primary open-angle glaucoma patients from the United States. *American journal of ophthalmology*, *155*(2), 342–53.e5.

Peterson, L. K., & Fujinami, R. S. (2007). Inflammation, demyelination, neurodegeneration and neuroprotection in the pathogenesis of multiple sclerosis. *Journal of neuroimmunology*, *184*(1–2), 37–44.

Piek, E., Sleumer, L. S., van Someren, E. P., Heuver, L., de Haan, J. R., de Grijs, I., . . . & van Zoelen, E. J. (2010). Osteo-transcriptomics of human mesenchymal stem cells: accelerated gene expression and osteoblast differentiation induced by vitamin D reveals c-MYC as an enhancer of BMP2-induced osteogenesis. *Bone*, *46*(3), 613–27.

Qureshi, I. A., & Mehler, M. F. (2012). Emerging roles of non-coding RNAs in brain evolution, development, plasticity and disease. *Nature reviews neuroscience*, *13*(8), 528–41.

Qureshi, I. A., & Mehler, M. F. (2013). Long non-coding RNAs: novel targets for nervous system disease diagnosis and therapy. *Neurotherapeutics*, *10*(4), 632–46.

Rezaie, T., Child, A., Hitchings, R., Brice, G., Miller, L., Coca-Prados, M., … & Sarfarazi, M. (2002). Adult-onset primary open-angle glaucoma caused by mutations in optineurin. *Science (New York, N.Y.)*, *295*(5557), 1077–79.

Rinn, J. L., Kertesz, M., Wang, J. K., Squazzo, S. L., Xu, X., Brugmann, S. A., . . . & Chang, H. Y. (2007). Functional demarcation of active and silent chromatin domains in human HOX loci by noncoding RNAs. *Cell*, *129*(7), 1311–23.

Riva, P., Ratti, A., & Venturin, M. (2016). The Long Non-Coding RNAs in Neurodegenerative Diseases: Novel Mechanisms of Pathogenesis. *Current Alzheimer research*, *13*(11), 1219–31.

Saccone, D., Asani, F., & Bornman, L. (2015). Regulation of the vitamin D receptor gene by environment, genetics and epigenetics. *Gene*, *561*(2), 171–80.

Santoro, M., Nociti, V., Lucchini, M., De Fino, C., Losavio, F. A., & Mirabella, M. (2016). Expression Profile of Long Non-Coding RNAs in Serum of Patients with Multiple Sclerosis. *Journal of molecular neuroscience: MN*, *59*(1), 18–23.

Sarfarazi, M., & Rezaie, T. (2003). Optineurin in primary open angle glaucoma. *Ophthalmology clinics of North America*, *16*(4), 529–41.

Sayad, A., Ghafouri-Fard, S., Omrani, M. D., Noroozi, R., & Taheri, M. (2017). Myxovirus resistance protein A (MxA) polymorphism is associated with IFNβ response in Iranian multiple sclerosis patients. *Neurological sciences*, *38*(6), 1093–99.

Shi, Q., & Yang, X. (2016). Circulating MicroRNA and Long Noncoding RNA as Biomarkers of Cardiovascular Diseases. *Journal of cellular physiology*, *231*(4), 751–55.

Sigdel, K. R., Cheng, A., Wang, Y., Duan, L., & Zhang, Y. (2015). The emerging functions of long noncoding RNA in immune cells: autoimmune diseases. *Journal of immunology research*, *2015*, 848790.

Soudyab, M., Iranpour, M., & Ghafouri-Fard, S. (2016). The Role of Long Non-Coding RNAs in Breast Cancer. *Archives of Iranian medicine*, *19*(7), 508–17.

Spurlock, C. F., 3rd, Tossberg, J. T., Guo, Y., Collier, S. P., Crooke, P. S., 3rd, & Aune, T. M. (2015). Expression and functions of long noncoding RNAs during human T helper cell differentiation. *Nature communications*, *6*, 6932.

Sundqvist, E., Bäärnhielm, M., Alfredsson, L., Hillert, J., Olsson, T., & Kockum, I. (2010). Confirmation of association between multiple sclerosis and CYP27B1. *European journal of human genetics*, *18*(12), 1349–52.

Taheri, M., Ghafouri-Fard, S., Solgi, G., Sayad, A., Mazdeh, M., & Omrani, M. D. (2017). Determination of cytokine levels in multiple sclerosis patients and their relevance with patients' response to Cinnovex. *Cytokine*, *96*, 138–43.

Tan, L., Yu, J. T., Hu, N., & Tan, L. (2013). Non-coding RNAs in Alzheimer's disease. *Molecular neurobiology*, *47*(1), 382–93.

Vučićević, D., Schrewe, H., & Orom, U. A. (2014). Molecular mechanisms of long ncRNAs in neurological disorders. *Frontiers in genetics*, *5*, 48.

Wan, P., Su, W., & Zhuo, Y. (2017). The Role of Long Noncoding RNAs in Neurodegenerative Diseases. *Molecular neurobiology*, *54*(3), 2012–21.

Wang, D. Q., Fu, P., Yao, C., Zhu, L. S., Hou, T. Y., Chen, J. G., . . . & Zhu, L. Q. (2018). Long Non-coding RNAs, Novel Culprits, or Bodyguards in Neurodegenerative Diseases. *Molecular therapy. Nucleic acids*, *10*, 269–76.

Wang, P., Xue, Y., Han, Y., Lin, L., Wu, C., Xu, S., . . . & Cao, X. (2014). The STAT3-binding long noncoding RNA lnc-DC controls human dendritic cell differentiation. *Science*, *344*(6181), 310–13.

Weber, F., Fontaine, B., Cournu-Rebeix, I., Kroner, A., Knop, M., Lutz, S., . . . & Müller-Myhsok, B. (2008). IL2RA and IL7RA genes confer susceptibility for multiple sclerosis in two independent European populations. *Genes and immunity*, *9*(3), 259–63.

Wu, G. C., Pan, H. F., Leng, R. X., Wang, D. G., Li, X. P., Li, X. M., & Ye, D. Q. (2015). Emerging role of long noncoding RNAs in autoimmune diseases. *Autoimmunity reviews*, *14*(9), 798–805.

Wu, P., Zuo, X., Deng, H., Liu, X., Liu, L., & Ji, A. (2013). Roles of long noncoding RNAs in brain development, functional diversification and neurodegenerative diseases. *Brain research bulletin*, *97*, 69–80.

Yap, K. L., Li, S., Muñoz-Cabello, A. M., Raguz, S., Zeng, L., Mujtaba, S., . . . & Zhou, M. M. (2010). Molecular interplay of the noncoding RNA ANRIL and methylated histone H3 lysine 27 by polycomb CBX7 in transcriptional silencing of INK4a. *Molecular cell*, *38*(5), 662–74.

Ying, L., Huang, Y., Chen, H., Wang, Y., Xia, L., Chen, Y., . . . & Qiu, F. (2013). Downregulated MEG3 activates autophagy and increases cell proliferation in bladder cancer. *Molecular biosystems*, *9*(3), 407–11.

Zahoor, I., & Haq, E. (2017). Vitamin D and multiple sclerosis: an update. In I. S. Zagon (Eds.) et. al., *Multiple Sclerosis: Perspectives in Treatment and Pathogenesis*. Codon Publications.

Zhang, F., Gao, C., Ma, X. F., Peng, X. L., Zhang, R. X., Kong, D. X., . . . & Hao, J. W. (2016). Expression profile of long noncoding RNAs in peripheral blood mononuclear cells from multiple sclerosis patients. *CNS neuroscience & therapeutics*, *22*(4), 298–305.

Zhang, Y., & Cao, X. (2016). Long noncoding RNAs in innate immunity. *Cellular & molecular immunology*, *13*(2), 138–47.

CHAPTER 4

LONG NON-CODING RNAS IN INFECTION AND IMMUNITY

QAMAR TABAN,[1,2]
PEERZADA TAJAMUL MUMTAZ,[1,3]
SAMEENA ISMAIL,[4] EHTISHAMUL HAQ,[1]
MASHOOQ DAR,[1] INSHA ZAHOOR,[2]
AND SYED MUDASIR AHMAD*[1]

[1] Division of Animal Biotechnology, Faculty of Veterinary Sciences and Animal Husbandry, Alusteng, Shuhama, SKUAST-Kashmir-190006, India
[2] Drug Therapeutics and Neurobiology Lab, Department of Biotechnology, School of Biological Sciences, University of Kashmir, Hazratbal, Srinagar, Jammu and Kashmir-190006, India
[3] Department of Biochemistry, School of Life sciences, Jaipur National University, Jaipur, Rajasthan-302017, India
[4] Department of Biotechnology, Government Degree College Anantnag, Jammu and Kashmir-192101
*Corresponding author
Syed Mudasir Ahmad
Division of Animal Biotechnology, Faculty of Veterinary Sciences and Animal Husbandry, SKUAST-K
Tel: 9419514131; E-mail: mudasirbio@gmail.com

Abstract

Immune responses combat various infectious agents by inducing inflammatory responses, antimicrobial pathways, and adaptive immunity. The polygenic responses to these external stimuli are regulated temporally and in coordination. Specific lncRNAs are induced to modulate innate and adaptive immune responses, which can function through various target interactions, like RNA-DNA, RNA-RNA, and RNA-protein interaction and hence affect the immunogenic regulation at various stages of gene expression. LncRNA are found to be present in various immune cells, like monocytes, macrophages, dendritic cells, neutrophils, T cells, and B cells. They have been shown to be involved in many biological processes, including the regulation of the expression of genes, dosage compensation, and genomics imprinting, but our understanding of how lncRNAs are regulated and how they alter cell differentiation/function is still obscure. Further dysregulation of lncRNA has been seen in many diseases; however, as yet, very little research has been carried out to understand the role of lncRNAs in regulation during host–pathogen interactions. In this review, we summarize the functional developments and mechanism of the action of lncRNAs, in immunity and defense of a host against pathogens.

Keywords: lncRNA, immunity, host–pathogen interaction, bacteria, viruses, autophagy

Introduction

In order to limit the autoimmune response to self-antigens and focus on immunity against non-self-antigens, the mechanisms of the immune system for gene expression regulation are crucial. The functional aspect of RNA is relatively less well known compared with the versatility of protein functions, which has been the main focus. While RNA molecules are traditionally known to act as passive carriers of information, the active participant in regulating gene expression at all levels is the RNA. Advancement in transcriptome sequencing has revealed that more than half the human genome is non-coding RNA (ncRNA) and a majority is constituted by long non-coding RNAs (lncRNAs), which are 200nt in length. The functional properties of lncRNAs in infections and immune cells are poorly understood, but during innate immune responses, many lncRNAs play a pivotal role. LncRNAs have recently been reported to have regulatory functions as guides for chromatin remodeling complexes, through interactions with transcriptional factors by acting as signals both in a cis and in a trans manner, and as scaffolds for many proteins; others have been shown to interfere with the transcriptional machinery or maintain the structure of nuclear speckles (Prasanth et al. 2005; Clemson et al. 2009; Sunwoo et al. 2009). Furthermore, lncRNAs also act post-transcriptionally in mRNA decay, as regulators of the splicing process, protein translation, and protein stability, and as molecular decoys for microRNAs (Yoon et al. 2013; Quinn and Chang 2016). LncRNAs play an important role in T-cell development and differentiation and activate both the innate and acquired branches of immunity. LncRNAs are found to be abundantly present in both innate and acquired immune cells and play an important role in gene expression regulation. Unraveling the highly heterogeneous nature of lncRNAs has revealed their versatile functions to observe variable molecular interactions thus imparting gene expression regulation and directing host–pathogen interactions.

LncRNAs in hematopoietic development

Productive immune response demands self-renewal and differentiation of hematopoietic stem cells (HSCs) into specialized blood lineage cells. These abilities are tightly controlled through the coordination between cell-type-specific genes, which suggests that lncRNAs might act as intermediates between extrinsic and intracellular signaling. Long-term HSCs show a high expression of the lncRNA H19. However, the lncRNA expression is down-regulated when long-term HSCs differentiate into short-term HSCs. Thus, a critical role is served by this lncRNA in sustaining long-term hematopoietic quiescence, because the depletion of maternal H19 during hematopoiesis is known to be associated with a gain in short-term HSCs (Venkatraman et al. 2013). The transcriptome of purified HSCs in a study identified unannotated lncRNAs through deep sequencing technology. Two hematopoietic stem cell lncRNAs were functionally assessed by knockdown studies: lncHSC-1 and lncHSC-2. The aim of this was to determine the extent of lncRNAs' involvement in hematopoietic differentiation. These two lncRNAs, LncHSC-1 and LncHSC-2, are principally located in the nucleus and differentially expressed between WT and Dnmt3a KO HSCs. Depletion of lncHSC-1 leads to the development of altered myeloid differentiation (MyD); further, the attenuation of lncHSC-2 is involved in impaired HSC self-renewal and increased T-cell differentiation. Enhancer regions that are marked by histone H3 Lys27 acetylation (H3K27ac) or histone H3 Lys4 monomethylation (H3K4me1) but not H3K4me3 and H3K27me3 are involved in the transcription of LncHSC-1 and LncHSC-2. LncHSC-1 is located near two coding genes, Zfp36l2 and Thada. LncHSC-2 binding sites are specific to the hematopoietic differentiation-specific transcriptional factor E2A; however, whether LncHSC-2 binds directly to TFs or recruits them through other complexes needs to be further examined (Luo et al. 2015).

LncRNAs in myeloid differentiation

In the bone marrow, short-lived myeloid cells, including macrophages and dendritic cells, are derived from common myeloid progenitor cells through various MyD88 pathways and it is suggested that lncRNAs have critical roles along these diverse pathways. Classical DCs (cDCs) or myeloid DCs as professional antigen-presenting cells (APCs) act as a link between innate and adaptive branches of immunity. On the other hand, Plasmacytoid DCs (pDCs) produce high levels of type I IFNs during viral and bacterial infections. A group of lncRNAs is differentially expressed during the differentiation of monocytes into human dendritic cells (DCs) as known through transcriptome microarray analysis and RNA-Seq.

Lnc-DC

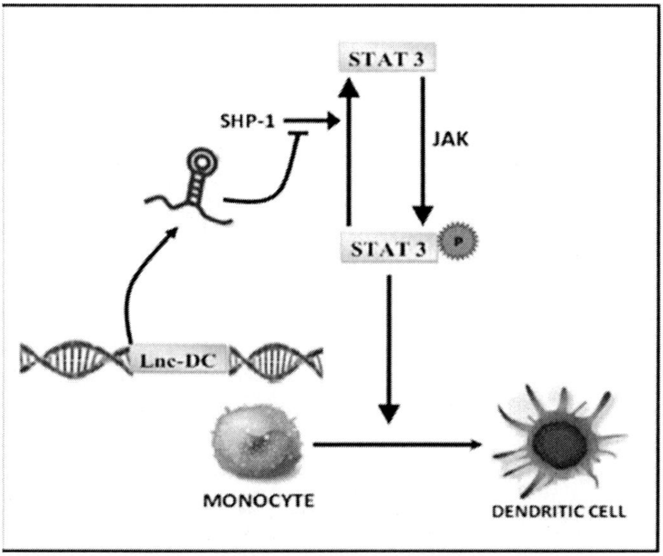

Figure 1. Lnc-DC promotes STAT3 phosphorylation for the differentiation of monocytes into dendritic cells. (Adopted from P. T. Mumtaz et al., 2017, *Biological procedures online*.)

Lnc-DC contributes to DC differentiation and is found to be up-regulated during human DC development (Wang et al. 2014). It is highly expressed in Lin – MHC II + CD11c + conventional DCs, but absent from pDCs, monocytes, and other leukocyte subsets. Lnc-DC also contributes to the regulating transcriptional factor STAT3 (signal transducer and activator of transcription 3) (figure 1). In the cytoplasm, it binds and blocks STAT3 and results in the maintenance of the phospho-STAT3 state and DC gene expression. During human DC development, the knockdown of lnc-DC has broad functional consequences, such as an altered expression of cell-surface receptors like CD80/86, CD40, and HLA-DR; an altered antigen uptake by monocyte-derived DCs; and decreased IL-12 production after LPS stimulation. A promoter-binding transcriptional factor, PU.1, is known to regulate lnc-DC transcription, which suggests its regulatory role in gene expression. Furthermore, activation of histone modifications by H3K4me3 and H3K27ac on the loci of lnc-DC thereby forms an accessible complex structure in human DCs (Wang et al. 2014). Lnc-DC also encodes a small secreted protein called Wdnm1-like (Dijkstra and Ballingall 2014) in mice; therefore, to better understand the role of lnc-DC in murine DC differentiation, further studies are required.

Morrbid

The lncRNA Morrbid (myeloid RNA regulator of Bim-induced death) regulates the lifespan of myeloid cells, for example, neutrophils, eosinophils, and monocytes, in both mice and humans in response to cytokines. Morrbid is conserved across many species containing five exons, it is poly-adenylated and is predominately localized to the nucleus bound to chromatin. As Bim is located downstream of Morrbid, it is thus known to regulate the myeloid cell lifespan. In Morrbid-deficient mice, the protein and mRNA levels of Bcl2l11 (Bim) were markedly elevated suggesting that Morrbid represses Bcl2l11 expression by promoting the deposition of H3K27me3 at the bivalent promoter of Bcl2l11. Loss of Morrbid in hematopoietic cells is shown to decrease the number of short-lived myeloid cells due to an increase in the death rate of terminal cells (Kotzin et al. 2016). In humans, Morrbid levels are dysregulated in patients with hypereosinophilic syndrome. Pro-survival cytokines can also potently influence the lifespan of immune cells through the repression of Bcl2l11 (Shinjyo et al. 2001). Cytokines from the common β-chain receptor family (IL-3, IL-5, and GM-CSF), which are known to promote the survival of eosinophils, neutrophils, and Ly6Chi monocytes, regulate Bcl2l11 expression in a Morrbid-dependent manner.

HOTAIRM1

HOTAIRM1 (HOXA transcript, antisense RNA myeloid-specific 1), an antisense lncRNA, is implicated in myeloid cell development. It is up-regulated in retinoic acid-induced normal human HSCs. HOTAIRM1 controls the expression of two neighboring genes that are involved in the transcriptional regulation of acute myeloid leukemia (AML) (Eklund 2006; Rice and Licht 2007) and normal hematopoiesis (Bei et al. 2007), that is, HOXA1 and HOXA2. It is also known to control CD11b and CD18 expression, which is crucial for myeloid cell differentiation (Zhang et al. 2009). The knockdown of HOTAIRM1 in the NB4 acute promyelocytic leukemia cell line blunts the retinoic acid-induced expression of HOXA1 and HOXA2 (but not distal HOXA genes) as well as CD11b and CD18 genes, which results in retarding all-trans retinoic acid (ATRA)-induced granulocytic differentiation and larger populations of immature and proliferating cells.

LncRNAs in cd4+ T-cell differentiation

LncRNAs are known to be the critical molecular determinants in the initiation of pathogen-specific adaptive immunity and are particularly involved in the differentiation into helper T cells. Helper T cells orchestrate adaptive immunity and are involved in the pathogenesis of various immunological diseases like autoimmunity and so on (Zhu et al. 2009).

LincR-CCR2-5′AS is expressed in the TH2 subset of helper cells and its expression is regulated by GATA-3 (Hu et al. 2013). LincR-CCR2-5′AS is known to up-regulate the cluster of genes that encode the key TH2 cell chemokines CCR1, CCR2, CCR3, and CCR5. The expression of LincR-Ccr2-5′AS is lowered resulting in the dysregulation of many genes including several chemokine receptor-encoding genes on GATA-3 shRNA silencing. In mice LincR-CCR2-5′AS is also known to regulate the migration of TH2 cells to the lung tissue *in vivo* (Yagi et al. 2011). Transcriptional factors STAT4 and STAT6 are known to regulate the differentiation of TH1 and TH2 cells, respectively (Vahedi et al. 2012). The majority of the lincRNA-encoding gene expression is regulated by STAT4 in TH1 cells, as assessed through comparison with STAT4-deficient and wild type cells. Similarly, STAT6 binds to and mediates the activation of the majority of lincRNA-encoding clusters including LincREpas1-3′AS in TH2 cells (Wei et al. 2010). Other than STAT proteins, many transcriptional factors have been linked to T-cell differentiation, such as T-bet for the differentiation of TH1 cells (Szabo et al. 2000). T-bet binds to lincRNA-encoding gene clusters in TH1 cells, including LincR-Ifng-3′AS and preferentially is known to contribute to positive expression in TH1 cells. The novel lncRNA TH2-LCR, which is located at the 3′ end of the RAD50 gene, has been identified. The lncRNA TH2-LCR is known for its regulatory role in transcription of the gene-cluster encoding TH2 cell cytokines, IL4, IL5, and IL13, which are expressed in human TH2 cells and the depletion of all TH2-LCR lncRNA alternatively spliced transcripts, resulting in a profound loss of the expression of cytokines, IL-4, IL-5, and IL-13. On the other hand, in murine T cells, the TH2-LCR is necessary to establish histone H3K4Me marks in IL4, IL13, and IL5 promoters and the distal DNA regulatory elements surrounding these genes. One function of lncRNAs is to recruit the enzymes required to produce H3K4Me marks in the genomic loci, of which each TH2-LCR lncRNA transcript was highly abundant in the nuclear fraction (Spurlock et al. 2015). In TH1 cells, the linc-MAF-4 chromatin-associated CD4+ lincRNA is known to repress expression of the TH2 cell transcription factor c-MAF4 through epigenetic silencing, which aids in T-cell differentiation toward the TH1 cell lineage. The kinetics of lincRNA-MAF-4 and c-MAF expression are inversely correlated in TH1 and TH2 cells *in vitro*, suggesting potential functional roles in the development of these helper T cells. It has been demonstrated that there is an interaction between linc-MAF-4 and MAF and the former might act as a scaffold to recruit two chromatin modifiers EZH2 and LSD1 to modulate the enzymatic activity of EZH2 on the MAF promoter and thus regulate its transcription (Ranzani et al. 2015).

LncRNAs in the activation of inflammation

LncRNAs are known to regulate the proinflammatory response or cytokine release. The innate immune response, being the frontline defense response of a cell, is important if the host is to defend itself from the invasion of various pathogenic organisms. Toll-like receptors (TLRs) are the innate immune system–associated specific receptors that help in eliminating a wide range of pathogens and are found to be the major receptor type of innate immunity for pathogen recognition (Turner et al. 2014). As cells encounter pathogens, an innate immune response is triggered by the production of inflammatory mediators or cytokines by activation of different types of transcription factors, such as API, NF-kB, and so on, via various TLR signaling pathways, for example, the toll-like receptor 4 (TLR4) bipartite signaling pathway (Rossetto et al. 2013). Mononuclear phagocytes represent the first line of defense against invading pathogens and macrophages being the key component of host defense. Many up-regulated lncRNAs in macrophages are associated with the toll-like receptor 2 (TLR2) ligands (Sonkoly et al. 2008). An inflammatory response is linked to lincRNA-Cox2, which is located upstream of the cyclooxygenase 2 gene (COX2, also known as prostaglandin endoperoxide synthase 2 or PTGS2). The expression of lincRNA-Cox2 is induced following activation of the TLR1 and TLR2 or TLR7 and TLR8, but not after activation of TLR3, in bone-marrow-derived DCs and macrophages challenged with LPS, Pam3CSK4, and R848 (Fig. 2). The expression is remarkably induced through the receptor adaptor MyD88 and the transcription factor NF-κB

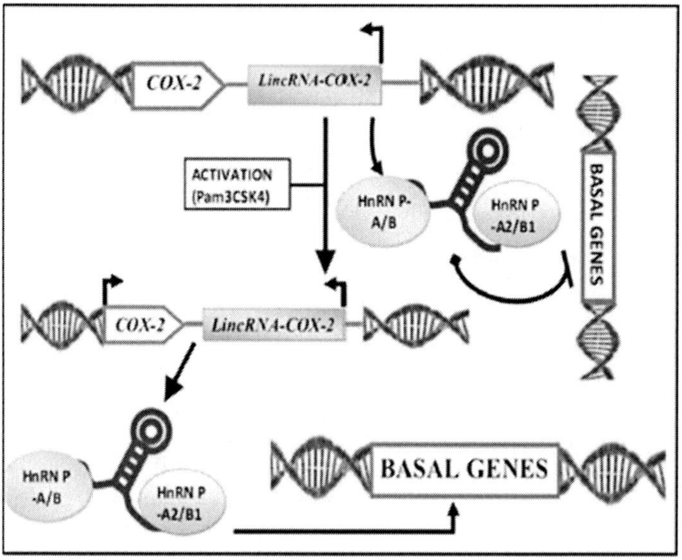

Figure 2. LincRNA-COX-2 is the localized 3′ of the COX2 gene, it effects expression of inflammatory genes and transcription repression of anti-inflammatory genes in non-stimulated cells and enhances the expression of proinflammatory genes on exposure of Pam3Csk4 through an interaction with hnRNP-A/B and hnRNP-A1/B1. (Adopted from P. T. Mumtaz et al., 2017, *Biological procedures online*.)

(Guttman et al. 2009; Carpenter et al. 2013). In resting macrophages, lincRNA-Cox2 regulates the gene expression of TLR1, IL6, and IL23a, chemokine receptors (CCRl), chemokines (CCL5 and CX3CL1), and interferon-stimulated genes like IRF7, OAS1a, OAS1l, and ISG15. Transcriptional repressive functions for the regulation of genes by lincRNA-Cox2 take place through the proposed mechanism of its interactions with the heterogeneous nuclear ribonucleoproteins hnRNP-A/B and hnRNP-A2/B1. Heterogeneous nuclear ribonucleoproteins (hnRNPs) are large nuclear RNA-binding proteins; depletion of these in macrophages dysregulates a set of genes encoding inflammatory molecules that overlap with the set of genes dysregulated by the knockdown of lincRNA-Cox2, suggesting a cooperative function for these molecules. For instance, lincRNA Cox2 plays a role in the TLR-induced expression of the IL-6 gene (Kawai and Akira 2010). It modulates cytokine genes and acts as a novel epigenetic regulator of both macrophage and intestinal epithelial inflammatory responses. In addition, it also regulates downstream of the NF-κB gene in epithelial cells. In intestinal epithelial cells exposed to TNF-α, the knockdown of lincRNA-Cox2 results in the reprogramming of the gene expression profile—in particular, transcriptional repression of IL-12b. This function is mediated via its interactions with Mi-2 nucleosome remodeling and the deacetylase (Mi- 2/NuRD) repressor complex, which this lincRNA appears to guide to the Il12b promoter region (Tong et al. 2016). All these studies agree that lincRNA-Cox2 has a broad acting role and unprecedented effects by facilitating the induction of a distinct cluster of immune response genes, thus playing an important role in inflammation.

LncRNA THRIL

LncRNA THRIL (TNF and heterogeneous nuclear ribonucleoprotein L related immunoregulatory lincRNA) is another regulator of TLR signaling that also functions through RNA–protein interactions with an hnRNP discovered via microarray analysis of the activated THP1 monocyte. Similar to the function of lincRNA-Cox2, THRIL also regulates the expression of inflammatory genes at a steady state including CXCL10, CCL1, CSF1, TNF, and IL8, suggesting its role in immune regulation. The expression of THRIL is involved with inflammation in Kawasaki disease as well (Li et al. 2014). However, the expression of THRIL is significantly decreased among recently discovered differentially expressed lncRNAs in response to stimulation by Pam3CSK, which is a TLR2 ligand discovered in a study using a differentiated human macrophage-like THP1 cell model. THRIL is also found to interact with heterogeneous nuclear ribonucleoprotein L (hnRNPL); the THRIL-hnRNPL complex binds to the TNFα promoter, thereby regulating its transcription in both basal and Pam3CSK4-activated conditions (figure 3). The specific interaction between THRIL and hnRNPL decreases the production of the cytokine TNF through macrophages, suggesting its effect on TLR-stimulated gene transcription. Loss-of-function studies revealed that THRIL promotes the inducible expression of the proinflammatory cytokine mediators TNF-α and IL-6 upon Pam3CSK4 stimulation (Mao et al. 2015). Thus, THRIL acts as a novel negative feedback regulator for the termination of TNFα expression, which marks the significant function of the lncRNA immune-related gene expression (Imamura and Akimitsu 2014).

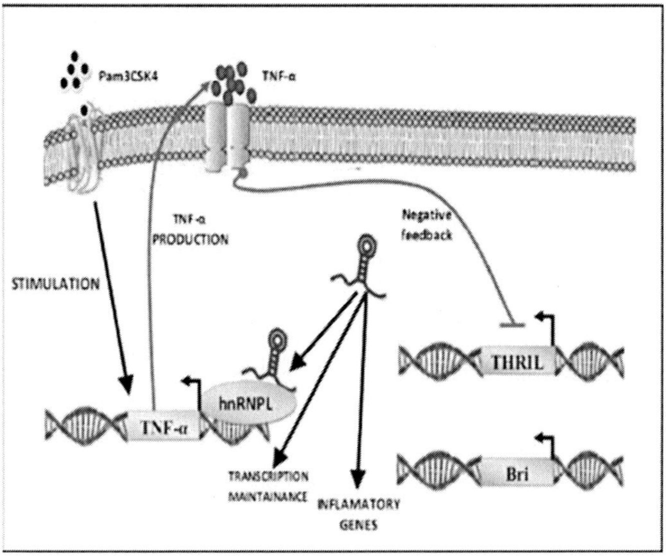

Figure 3. THRIL was identified as an antisense lncRNA that enhances the transcription of TNF by constituting a complex with hnRNPL and binding to the promoter of TNF. THRIL also has a regulatory role in basal and Pam3Csk4-induced gene expression. (Adopted from P. T. Mumtaz et al., 2017, *Biological procedures online*.)

LncRNA PACER

LncRNA PACER (p50-associated Cox2 extragenic RNA) is a well-known lncRNA that is located upstream of the COX-2 transcriptional start site and is expressed in the antisense direction in humans (figure 4). PACER plays a regulatory role in the gene expression of COX-2 (cyclooxygenase 2'), a known mediator of the inflammatory response. PACER is induced by stimulation with bacterial LPS, and thus specifically regulates its neighboring gene PTGS2, which encodes COX-2 (Smith et al. 2000). It acts as a decoy for the NF-κB signaling pathway by binding the NF-κB subunit p50 and is involved in sequestering the transcriptional factor away from the PTGS2 promoter and of other target genes (Krawczyk and Emerson 2014). This sequestration facilitates the recruitment of histone acetyltransferase (HAT) p300 and the assembly of the RNA polymerase II pre-initiation complex at the promoter of the COX2 gene. PACER expression is induced by the chromatin boundary/insulator factor CCCTC-binding factor (CTCF), which in turn forms a conducive chromatin environment in the upstream region of the COX2 gene (Krawczyk and Emerson 2014). The PACER homolog in mice has been identified as cyclooxygenase II enzyme-divergent (Ptgs2os), whose expression in mouse embryonic fibroblasts (MEFs) is highly induced by LPS, IL-1β, and TNF and various TLR agonists viz., Pam3CK4, HKLM, and Poly (I:C). Interestingly, the Cox2-divergent displays similar up-regulated expression patterns upon cytokine/TLR agonist stimulations in RelA/MEFs, as compared with wild type MEFs, suggesting the indirect regulation of lncRNA Cox2-divergent by RelA (NF-κB component) (Rapicavoli et al. 2013). The LncRNA Cox2-divergent homolog is expressed in primary human mammary epithelial cells (pHMECs) and in human monocytes that are involved in the processes like macrophage differentiation (Krawczyk and Emerson 2014). Overall, the involvement of the PACER lncRNA in multiple processes related to the regulation of immunogene expression has been highlighted. Moreover, lncRNA lnc-IL7R is found to have immune system–associated functions as well.

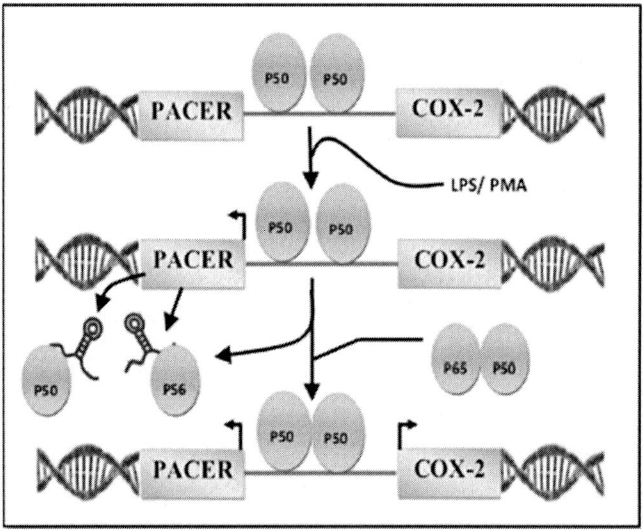

Figure 4. PACER is localized upstream of the Cox2 transcriptional start site. It induces Cox2 expression by inhibiting the repressive action of the p50 homodimer (of NF-kB) bound at the Cox2 promoter site. (Adopted from P. T. Mumtaz et al., 2017 *Biological procedures online*.)

Lnc-IL7R

A novel lncRNA viz, lnc-IL7R is identified and is up-regulated in LPS-stimulated human monocytic THP-1 cells and is transcribed from the 3′ UTR of IL-7R gene in its sense orientation. Also, the expression of lnc-IL7R was found to be up-regulated in human peripheral blood mononuclear cells (PBMNC). Lnc-IL7R is known to down-regulate the expression of IL-6, IL-7R, IL-8, VCAM-1, and E-selectin. Furthermore, the knockdown study of lnc-IL7R revealed a decrease in the trimethylation of histone H3K27 at promoters of inflammatory mediators, suggesting that lnc-IL7R epigenetically regulates inflammatory responses (Cui et al. 2014).

LncRNAs in restricting inflammatory responses

Many studies have identified protective functions for lncRNAs by restricting excessive inflammatory responses.

LncRNA Lethe

In induced mouse embryonic fibroblasts, lncRNA Lethe, a Rps15a pseudogene (Rps15a-ps4), was first identified as a functional pseudogene following stimulation with IL-1β and TNF. Lethe has recently been revealed to be localized in chromatin and is believed to function as a negative regulator of NF-κB by binding to RelA (p65), resulting in the inhibition of RelA, thence regulating the NF-κB target gene expressions, such as IL-8, IL-6, and SOD2. The knockdown of Lethe results in the up-regulation of NF-κB targets, which suggests a negative regulatory function, as an induced expression of Lethe is dependent on NF-κB activity. Lethe is markedly up-regulated in response to stimulation, with glucocorticoid receptor agonists such as dexamethasone, cytokines like IL-1β, and TNF-α; however, the expression of Lethe is not responsive to TLR agonist challenges (Rapicavoli et al., 2013). Hence, Lethe acts as a decoy receptor for NF-κB and provides negative feedback in response to immune system activation.

LncRNA lincRNA-EPS

LncRNA lincRNA-EPS is expressed in erythrocytes and in macrophages and DCs; its expression is down-regulated after activation of the innate immune system. Macrophages generated from lincRNA-EPS-deficient mice that are resting or are stimulated with TLR ligands show a higher expression of genes encoding IL6, CXCL10, CCL4, and IRG1 than do their wild counterparts. After stimulation of the immune system, lincRNA-EPS-deficient macrophages displaying epigenomic changes are lost. LincRNA-EPS acts directly on chromatin to restrain the expression of genes by controlling chromatin accessibility and nucleosome positioning in immune cells (Atianand et al. 2016).

Lnc13

Lnc13 is another lncRNA that localizes to the nucleus and functions to repress the transcription of immune-related genes in a manner similar to lincRNA-EPS. Lnc13 is expressed in macrophages and on activation by TLR4 is down-regulated. It functions in an RNA–protein complex with hnRNPD and the histone deacetylase HDAC1 to suppress the transcription of a distinct set of genes encoding immune-response molecules, including MyD88, STAT1, STAT3, and TNF (Castellanos-Rubio et al. 2016).

LncRNA NKILA

The lncRNA NKILA has been shown to negatively regulate NF-κB signaling, but at the level of post-transcriptional control, NKILA is substantially induced in breast cancer cells following stimulation with IL-1β or TNF. NKILA forms RNA–protein complex with NF-κB and its negative regulator IκB (inhibitor of NF-κB') and prevents phosphorylation motifs of IκB to bind IKK ("IκB kinase") (Liu et al. 2015).

LncRNAs in T-cell activation

LncRNA NeST

LncRNA NeST (Nettoie Salmonella pas Theiler's), also called Tmevpg1, located downstream adjacent to the IFN γ-encoding gene, is transcribed convergently to the IFN-γ gene in both humans and mice. The location and orientation of human NeST are conserved, but the primary transcript includes the opposite strand of the entire IFNγ gene. Congenic or transgenic expression of NeST increases IFN-γ production in activated CD8+ T cells (Gomez et al. 2013). NeST is expressed in CD8+ T cells, CD4+ TH1 cells and natural killer cells; its expression depends on the combinatorial activity of the transcription factors T-bet, STAT4, and NF-κB (Collier et al. 2012; Vigneau et al. 2003). These studies identify a critical role in T-cell function and in the systems-level immune response to viral infection *in vivo*. Mice over-expressing NeST show increased susceptibility to Theiler's virus in terms of persistence but remarkable resistance to Salmonella pathogenesis. Mechanistic analysis indicates its interaction with WDR5, a core subunit of the histone H3K4 methyltransferase complex, leading to the modification of H3 methylation at the IFN-γ locus, thereby epigenetically regulating IFN-γ expression. In developing and differentiated effector Th1 cells, T-bet guides epigenetic remodeling of lncRNA NeST distal and proximal enhancers leading to the recruitment of stimulus-inducible transcription factors (e.g., NF-κB and Ets-1) to the locus to achieve the Th1- lineage-specific expression of IFN-γ (Collier et al. 2014).

LncRNAs in restricting excessive activation of T cells

NRON

NRON is a non-coding repressor of NFAT (Nuclear Factor of Activated T cells), first identified during a short hairpin RNA (shRNA) library screening against 512 evolutionarily conserved lncRNAs (Willingham et al. 2005). NRON acts as a repressor for the activation of the calcium-dependent NFAT by sequestering the phosphorylated form of NFAT in the cytoplasm. NFAT is a highly phosphorylated transcriptional factor present in the cytoplasm of resting cells. NRON nucleates a large RNA–protein complex in resting T cells including the calmodulin-binding protein IQGAP1, the kinase LRRK2, the dual-specificity tyrosine phosphorylation-regulated kinase (DYRK), Casein kinase 1 (CK1), Glycogen synthase kinase 3 (GSK3) (Sharma et al. 2011), and the nuclear-transport factor karyopherin β1. After the stimulation of the T cell receptor, NFAT is released from the NRON complex and is dephosphorylated by the Ca2+-calmodulin-dependent phosphatase calcineurin and then is transported into the nucleus, where it can activate transcription and induce expression of target genes, such as IL-2 (figure 5). Accordingly, the depletion of NRON results in the increased dephosphorylation of NFAT, the translocation of NFAT to the nucleus, the nuclear accumulation of NFAT, and the production of cytokines following the activation of T cells. Thus, this suggests that NRON acts as a transcription repressor by inhibiting nucleocytoplasmic shuttling of NFAT and that NRON can function as a transcriptional regulator for immune regulation (Sharma et al. 2011).

LncRNA CD244

In CD8+ T cells, the lncRNA-CD244 restricts the activation of T cells. The expression of lncRNA-CD244 is induced by signaling through the T-cell inhibitory receptor CD244 (2B4) and mediates the repression of genes encoding the inflammatory molecules IFN-γ and TNF. Notably, the knockdown of lncRNA-CD244 improves the CD8+ T-cell function *in vivo*. In a model of infection with *mycobacterium tuberculosis*, the transfer of CD8+ T cells depleted of lncRNA-CD244 resulted in significantly lower organ bacterial burden in mice than did the transfer of lncRNA-CD244-sufficient (control) T cells (Wang et al. 2015).

LncRNA expression in B lymphocytes

B cells secrete specific antibodies, and provide long-lasting defensive immunity against pathogens. However, our understanding of lncRNAs in B cells is ambiguous. From several transcriptome-profiling studies carried out in mice (Brazao et al. 2016) and humans (Bonnal et al. 2015), the expression profiles of lncRNAs in B cells have been obtained. The transcription factor PAX5, crucial for B-cell development, acts as an important driver of lncRNA expression in pro-B and mature B cells in mice. LncRNA FAS-AS1 regulates FAS-receptor (CD95; TNFRSF6) signaling (Sehgal et al. 2014) and its expression is inversely correlated with sFAS-induced apoptosis in B-cell lymphomas by blocking the splicing of FAS pre-mRNA (Niitsu et al. 1999). LncRNAs are also involved in the activation and development of B cells. During the development of B cells in bone marrow, they acquire immunoglobulin B-cell receptors through the V(D)J recombination process. In peripheral lymphoid organs, activation of B cells during inflammation or infection is mediated through the enzyme AID (activation-induced deaminase), which controls two crucial processes, namely, somatic hypermutation and class switch recombination (Muramatsu et al. 2000). These processes are crucial for generating a vast repertoire of highly specific antibodies by B cells that regulate the humoral arm of adaptive immunity. LncRNA transcription (Bolland et al. 2004; Verma et al. 2012) and the exosome-dependent regulation of lncRNA abundance (Pefanis et al. 2015) are thought to regulate AID-mediated somatic hypermutation and class switch recombination.

LncRNAs in host defense against pathogens

The immune system plays an important role in defense against microbial pathogens. A class of host-encoded lncRNAs has been identified to play a functional role in controlling the host's immune responses upon microbial invasion. Some microbial species are also known to produce lncRNAs that play pivotal roles in pathogen life cycles as well as affecting host–pathogen interactions. The lncRNA-mediated regulation of host–pathogen interactions during pathogen invasion have also emerged that shall be discussed later.

PAN

Pathogens can also express functional lncRNAs that are known to affect host–pathogen interactions. One well-characterized pathogen/microbial-derived lncRNA is PAN RNA (polyadenylated nuclear RNA) (Rossetto and Pari 2014; Erdmann et al. 2001). The Kaposi's sarcoma–associated herpes virus (KSHV) genome encodes the PAN lncRNA where it is implicated to play important roles in the KSHV viral gene expression and replication processes (Borah et al. 2011). PAN interacts with Histone demethylases UTX and JMJD3, thereby recruiting histone-modifying complexes to the KSHV genome. Thus, PAN epigenetically regulates viral gene expression and viral life cycles by promoting the switch from latent to lytic infection (Rossetto et al. 2013). On the other hand, PAN RNA has a regulatory role in host immunity as well. The viral lncRNA PAN suppresses expression of host genes that are involved in the inflammatory and antiviral responses, including IFNγ, IL-18, IFNA16, and RNase L (Erdmann et al. 2001) (figure 5). A recent report showed that PAN can physically interact with polycomb group proteins, such as PRC2 and mediate repression of host-cell gene expression (Rossetto and Pari 2014). Taken together, PAN is a multifunctional viral lncRNA involved in the regulation of both viral and host gene expressions.

NRAV

The antisense lncRNA NRAV (negative regulator of antiviral response, also known as DYNLL1-AS1) is involved in antiviral innate immunity via a genome-wide profiling of lncRNA in influenza virus A/WSN/33 (H1N1)-infected human alveolar epithelial A549 cells. In humans, NRAV is a 1,177-base pair, polyadenylated and nuclear-localized RNA. The NRAV gene overlaps with the first intron of the protein-coding gene DYNLL1 (dynein light chain 1) and is transcribed in the antisense orientation. NRAV modulates virus replication, production, and virulence and is known to be down-regulated in human cells during various viral infections by numerous viruses, including ssRNA viruses such as influenza A (IAV), sendai virus (SeV), dsRNA viruses such as Muscovy Duck Reovirus (MDRV), and DNA viruses such as herpes simplex virus (HSV). The ectopic expression of NRAV in human or mouse cells during IAV replication and virulence suggests that the down-regulation of NRAV is an important component of the host antiviral immune system to efficiently control viral replication and clearance. NRAV represses several interferon-stimulated genes (ISGs), including IFITM3 and MxA, at their promoters (Ouyang et al. 2014) (figure 5).

LncRNAs and their involvement in viral infections

The analysis of lncRNA functions in viral infections pose unique challenges due to very low expression abundance, lack of sequence conservation, and limited annotation. But, many lncRNAs have been functionally implicated in viral infections. During viral infection, there is a multitude of changes throughout the transcriptome, which includes lncRNAs. Multiple methods for the detection of lncRNAs are currently available, but transcriptome deep sequencing technology (RNA-Seq) is likely the most preferred and widely used, in part due to its broad applicability and accessibility. LncRNAs are often involved in the host immune response and are virally-encoded, where both function epigenetically. Virally-encoded lncRNAs function epigenetically by interacting with DNMTs and chromatin-remodeling complexes. For example, lncRNA PAN epigenetically interacts with histone demethylases (HATs) that recruit histone-modifying complexes to the Kaposi's sarcoma–associated herpes virus genome and regulate gene expression and the viral life cycle (Rossetto et al. 2013). LncRNAs function as scaffolds in the formation of oddly-formed nuclear compartments in mammalian cells, known as paraspeckles.

NEAT1

NEAT1 (nuclear-enriched abundant transcript 1 or nuclear paraspeckle assembly transcript 1) was first identified as a nuclear lncRNA. It is an inducible lncRNA that was discovered in a mouse brain infected with Japanese encephalitis virus or rabies virus. NEAT1 has been shown to regulate HIV1 replication by transporting a newly synthesized Rev-dependent instability element (INS) containing HIV-1 mRNA from the nucleus to the cytoplasm (Zhang et al. 2013). It was also found that NEAT1 can be dramatically induced in HIV-1-infected T cells as well as influenza-virus- and herpes-simplex-virus-infected epithelial cells. NEAT1 expression is dependent upon the treatment of TLR3 ligand poly I:C, which mimics the effect of viral infection. NEAT1 is shown to bind to SFPQ (splicing factor proline- and glutamine-rich protein) and silence IL-8 expression via binding to the IL-8 promoter in normal physiological states. However, in response to viral infection, induction of NEAT1 results in the relocation of SFPQ from the IL-8 promoter to paraspeckles followed by triggering the transcriptional activation of IL-8 (Imamura et al. 2014) (figure 5). Taken together, lncRNA NEAT1 plays an important role in the innate immune response to viral infection.

Evidence suggests that virally-encoded lncRNAs also function as miRNA sponges. The interplay of virally-encoded lncRNAs and host miRNAs may prove to be an essential function of viral pathogenesis. A review describes the function of herpes virus Saimiri (HVS)-encoded lncRNAs that do have this function, HSUR1 and HSUR2, and highlights the susceptibility of host miRNAs in viral infections (Tavanez et al. 2014). The functional roles of miRNAs in HVS-transformed T cells, also revealed that miRNA-27 directly down-regulates several proteins in the T-cell receptor signaling pathway (Guo et al. 2015).

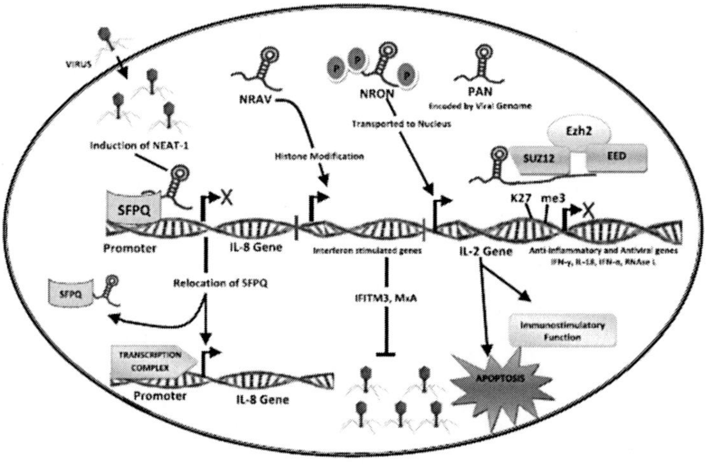

Figure 5. The viral lncRNA PAN suppresses expression of host genes involved in the inflammatory and antiviral responses, including IFNγ, IL-18, IFNA16, and RNase L. It can physically interact with polycomb group proteins, such as PRC2, and can mediate repression of host cellular gene expression. LncRNA NRAV also has a role in the initial transcription of multiple interferon-stimulated genes (ISGs), such as MxA and IFITM3, via epigenetically regulating histone modifications. NEAT1 binds to SFPQ (a paraspeckle protein) and plays an important role in the formation of the nuclear paraspeckle body, which on the other hand silences IL-8 expression via binding to the IL-8 promoter in normal physiological states. NFAT is a highly phosphorylated transcriptional factor present in the cytoplasm of resting cells. NFAT induces the expression of target genes, such as IL-2, which play a key role in enduring cell-mediated immunity. (Adopted from P. T. Mumtaz et al., 2017 *Biological procedures online*.)

Antisense LncRNAs in the regulation of cytokine genes

Antisense lncRNAs are known to have a restricted impact on gene expression, that is, to regulate specific complementary protein-coding transcripts. Situated within the IL-1β locus, a downstream eRNA, namely, IL1β-eRNA, and an upstream mRNA-surrounding RBT, called IL1β-RBT46, have been identified. IL1β-eRNA and IL1β-RBT46 are predominantly located in the nucleus of naive and LPS-stimulated cells; the expression of IL1β-eRNA and IL1β-RBT46 is mediated by NF-κB. The depletion of IL1β-eRNA and IL1β-RBT46 in human monocytes is known to selectively attenuate the LPS-induced expression of CXCL8 and IL-1β through unknown mechanisms (Ilott et al. 2014). The antisense-transcript of IL-1β (anti-IL1β transcript), IL1β- RBT46, and IL1β-eRNA are non-coding transcripts that are located close to the IL-1β gene. Anti-IL-1β is transcribed from the 5′ upstream promoter sequence of the coding gene IL-1β and in mouse macrophages; its ectopic over-expression significantly suppresses LPS-induced IL-1β expression in RAW264.7 cells. The antisense IL-1β seems to act as a natural antisense transcript of the IL-1β gene to regulate the homeostasis of IL-1β in cells. The anti-IL1β transcript modulates the chromatin structure surrounding the IL-1β promoter by decreasing H3K4 trimethylation (Lu et al. 2013).

Beyond IL-1β, the antisense IL-1α (AS IL-1α) RNA shows similar expression patterns to the IL-1α protein-coding gene and regulates the expression of IL-1α. Using shRNA approaches, AS IL-1α was shown to be essential for IL-1α gene transcription by compromising the recruitment of RNAPII to the promoter, thereby decreasing the levels of IL-1α mRNA in macrophages exposed to LPS (Chan et al. 2015). Like many inflammatory genes, a substantial number of lncRNAs are only expressed in innate immune cells following their activation, while other lncRNAs that are abundantly expressed are down-regulated when cells are exposed to inflammatory stimuli.

LncRNAs and their role in autophagy

In addition to the phagocytosis and secretion of pro-inflammatory mediators by macrophages, autophagy is a mechanism of the immune response to bacterial infection and maintains cellular homeostasis. Accumulating evidence has suggested that autophagy functions as a housekeeping, intracellular defense pathway in response to infection. Although the role of lncRNAs in the autophagic response to intracellular bacteria is still unknown, it has been reported that production of IFN- γ during mycobacterium bovis BCG infection induces autophagy in macrophages that result in the sustained down-regulation of lncRNA MEG3 maternally expressed gene 3 (MEG3), because lncRNA MEG3 inactivates mTOR. Thus, the knockdown of lncRNA MEG3 in macrophages results in autophagy and the enhanced removal of intracellular M. bovis BCG. Further research may enable us to better understand the role of lncRNAs in autophagy and may provide a new insight for effective therapeutics.

Conclusion and future perspectives

LncRNAs are the active participants in many processes such as multiple stages of immune-cell development and in pathogen-response pathways; however, functional annotations of many lncRNAs and their role in host–pathogen interactions still remain a mystery. Even with the fast increase in understanding of the functional perspectives and biological relevance of lncRNAs in the past decade, this field is still in its infancy and many questions and challenges are yet to be addressed. Efforts to recognize the role of lncRNAs as key regulators in host immune responses remain a challenge; nevertheless, multiple high-throughput methodologies coupled with computational strategies have paved the way for narrowing down the list of lncRNAs involved in host–pathogen interactions. RNA-directed technologies (e.g., ChIRP and RAP) would help uncover lncRNA interactomes, including protein partners and genomic-binding sites. Certainly, the advanced technologies and methodologies will continue to clarify how lncRNAs influence lncRNA-mediated regulation of host–pathogen interactions, and possibly some unique functions beyond the gene expression mechanism. In this chapter, we have discussed individual lncRNAs in various immune-related processes such as the development, differentiation, and effector functions of different classes of immune cells. Studying the mechanisms and continued investigation of the involvement of lncRNAs in host pathogenesis will pave the way for better therapeutic treatment of many infectious diseases related to humans.

Acknowledgement

We offer sincere gratitude to the Council of Scientific and Industrial Research (CSIR), Government of India, for financial support in the form of Junior and Senior Research Fellowships.

Conflict of interest

The authors declare that they have no conflicts of interest with respect to research, authorship, and publication of this book chapter.

Copyright and permission statement

We confirm that the materials included in this chapter do not violate copyright laws. Where relevant, appropriate permissions have been obtained from the original copyright holder(s). All original sources have been appropriately acknowledged and/or referenced.

Funding

No funding was received for this project.

References

Atianand, M. K., Hu, W., Satpathy, A. T., Shen, Y., Ricci, E. P., Alvarez-Dominguez, J. R., . . . & Braun, J. E. (2016). A long noncoding RNA lincRNA-EPS acts as a transcriptional brake to restrain inflammation. *Cell*, *165*(7), 1672–85.

Bei, L., Lu, Y., Bellis, S. L., Zhou, W., Horvath, E., & Eklund, E. A. (2007). Identification of a HoxA10 activation domain necessary for transcription of the gene encoding β3 integrin during myeloid differentiation. *Journal of biological chemistry*, *282*(23), 16846–59.

Bolland, D. J., Wood, A. L., Johnston, C. M., Bunting, S. F., Morgan, G., Chakalova, L., . . . & Corcoran, A. E. (2004). Antisense intergenic transcription in V (D) J recombination. *Nature immunology*, *5*(6), 630.

Bonnal, R. J., Ranzani, V., Arrigoni, A., Curti, S., Panzeri, I., Gruarin, P., . . . & Pagani, M. (2015). De novo transcriptome profiling of highly purified human lymphocytes primary cells. *Scientific data*, *2*, 150051.

Borah, S., Darricarrère, N., Darnell, A., Myoung, J., & Steitz, J. A. (2011). A viral nuclear noncoding RNA binds re-localized poly (A) binding protein and is required for late KSHV gene expression. *PLoS pathogens*, *7*(10), e1002300.

Carpenter, S., Aiello, D., Atianand, M. K., Ricci, E. P., Gandhi, P., Hall, L. L., . . . & O'Neill, L. A. (2013). A long noncoding RNA mediates both activation and repression of immune response genes. *Science*, *341*(6147), 789–92.

Castellanos-Rubio, A., Fernandez-Jimenez, N., Kratchmarov, R., Luo, X., Bhagat, G., Green, P. H., . . . & Ghosh, S. (2016). A long noncoding RNA associated with susceptibility to celiac disease. *Science*, *352*(6281), 91–95.

Chan, J., Atianand, M., Jiang, Z., Carpenter, S., Aiello, D., Elling, R., . . . & Caffrey, D. R. (2015). Cutting edge: a natural antisense transcript, AS-IL1α, controls inducible transcription of the proinflammatory cytokine IL-1α. *The journal of immunology*, *195*(4), 1359–63.

Clemson, C. M., Hutchinson, J. N., Sara, S. A., Ensminger, A. W., Fox, A. H., Chess, A., & Lawrence, J. B. (2009). An architectural role for a nuclear noncoding RNA: NEAT1 RNA is essential for the structure of paraspeckles. *Molecular cell*, *33*(6), 717–26.

Collier, S. P., Collins, P. L., Williams, C. L., Boothby, M. R., & Aune, T. M. (2012). Cutting edge: influence of Tmevpg1, a long intergenic noncoding RNA, on the expression of Ifng by Th1 cells. *Journal of immunology*, *189*(5), 2084–88.

Collier, S. P., Henderson, M. A., Tossberg, J. T., & Aune, T. M. (2014). Regulation of the Th1 genomic locus from Ifng through Tmevpg1 by T-bet. *Journal of immunology*, *193*(8), 3959–65.

Cui, H., Xie, N., Tan, Z., Banerjee, S., Thannickal, V. J., Abraham, E., & Liu, G. (2014). The human long noncoding RNA lncIL7R regulates the inflammatory response. *European journal of immunology*, *44*(7), 2085–95.

Dijkstra, J. M., & Ballingall, K. T. (2014). Non-human lnc-DC orthologs encode Wdnm1-like protein. *Research*, *3*.

Eklund, E. A. (2006). The role of HOX genes in myeloid leukemogenesis. *Current opinion in hematology*, *13*(2), 67–73.

Erdmann, V. A., Barciszewska, M. Z., Hochberg, A., De Groot, N., & Barciszewski, J. (2001). Regulatory RNAs. *Cellular and molecular life sciences*, *58*(7), 960–77.

Gomez, J. A., Wapinski, O. L., Yang, Y. W., Bureau, J. F., Gopinath, S., Monack, D. M., . . . & Kirkegaard, K. (2013). The NeST long ncRNA controls microbial susceptibility and epigenetic activation of the interferon-γ locus. *Cell*, *152*(4), 743–54.

Guo, Y. E., Oei, T., & Steitz, J. A. (2015). Herpesvirus saimiri MicroRNAs preferentially target host cell cycle regulators. *Journal of virology*, *89*(21), 10901–11.

Guttman, M., Amit, I., Garber, M., French, C., Lin, M. F., Feldser, D., . . . & Cabili, M. N. (2009). Chromatin signature reveals over a thousand highly conserved large non-coding RNAs in mammals. *Nature*, *458*(7235), 223.

Hu, G., Tang, Q., Sharma, S., Yu, F., Escobar, T. M., Muljo, S. A., . . . & Zhao, K. (2013). Expression and regulation of intergenic long noncoding RNAs during T cell development and differentiation. *Nature immunology*, *14*(11), 1190.

Ilott, N. E., Heward, J. A., Roux, B., Tsitsiou, E., Fenwick, P. S., Lenzi, L., . . . & Donnelly, L. E. (2014). Long non-coding RNAs and enhancer RNAs regulate the lipopolysaccharide-induced inflammatory response in human monocytes. *Nature communications*, *5*, 3979.

Imamura, K., & Akimitsu, N. (2014). Long non-coding RNAs involved in immune responses. *Frontiers in immunology*, *5*, 573.

Imamura, K., Imamachi, N., Akizuki, G., Kumakura, M., Kawaguchi, A., Nagata, K., . . . & Kai, C. (2014). Long noncoding RNA NEAT1-dependent SFPQ relocation from promoter region to paraspeckle mediates IL8 expression upon immune stimuli. *Molecular cell*, *53*(3), 393–406.

Kawai, T., & Akira, S. (2010). The role of pattern-recognition receptors in innate immunity: update on Toll-like receptors. *Nature immunology*, *11*(5), 373.

Kotzin, J. J., Spencer, S. P., McCright, S. J., Kumar, D. B. U., Collet, M. A., Mowel, W. K., . . . & Harman, C. C. (2016). The long non-coding RNA Morrbid regulates Bim and short-lived myeloid cell lifespan. *Nature*, *537*(7619), 239.

Krawczyk, M., & Emerson, B. M. (2014). p50-associated COX-2 extragenic RNA (PACER) activates COX-2 gene expression by occluding repressive NF-κB complexes. *Elife*, *3*.

Li, Z., Chao, T. C., Chang, K. Y., Lin, N., Patil, V. S., Shimizu, C., . . . & Rana, T. M. (2014). The long noncoding RNA THRIL regulates TNFα expression through its interaction with hnRNPL. *Proceedings of the National Academy of Sciences*, *111*(3), 1002–7.

Liu, B., Sun, L., Liu, Q., Gong, C., Yao, Y., Lv, X., . . . & Zeng, M. (2015). A cytoplasmic NF-κB interacting long noncoding RNA blocks IκB phosphorylation and suppresses breast cancer metastasis. *Cancer cell*, *27*(3), 370–81.

Lu, J., Wu, X., Hong, M., Tobias, P., & Han, J. (2013). A potential suppressive effect of natural antisense IL-1β RNA on lipopolysaccharide-induced IL-1β expression. *The journal of immunology*, *190*(12), 6570–78.

Luo, M., Jeong, M., Sun, D., Park, H. J., Rodriguez, B. A., Xia, Z., . . . & Li, W. (2015). Long non-coding RNAs control hematopoietic stem cell function. *Cell stem cell*, *16*(4), 426–38.

Mao, A. P., Shen, J., & Zuo, Z. (2015). Expression and regulation of long noncoding RNAs in TLR4 signaling in mouse macrophages. *BMC genomics*, *16*(1), 45.

Muramatsu, M., Kinoshita, K., Fagarasan, S., Yamada, S., Shinkai, Y., & Honjo, T. (2000). Class switch recombination and hypermutation require activation-induced cytidine deaminase (AID), a potential RNA editing enzyme. *Cell*, *102*(5), 553–63.

Niitsu, N., Sasaki, K., & Umeda, M. (1999). A high serum soluble Fas/APO-1 level is associated with a poor outcome of aggressive non-Hodgkin's lymphoma. *Leukemia*, *13*(9), 1434.

Ouyang, J., Zhu, X., Chen, Y., Wei, H., Chen, Q., Chi, X., . . . & Wang, G. (2014). NRAV, a long noncoding RNA, modulates antiviral responses through suppression of interferon-stimulated gene transcription. *Cell host & microbe*, *16*(5), 616–26.

Pefanis, E., Wang, J., Rothschild, G., Lim, J., Kazadi, D., Sun, J., . . . & Economides, A. N. (2015). RNA exosome-regulated long non-coding RNA transcription controls super-enhancer activity. *Cell*, *161*(4), 774–89.

Prasanth, K. V., Prasanth, S. G., Xuan, Z., Hearn, S., Freier, S. M., Bennett, C. F., . . . & Spector, D. L. (2005). Regulating gene expression through RNA nuclear retention. Cell, *123*(2), 249–63.

Quinn, J. J., & Chang, H. Y. (2016). Unique features of long non-coding RNA biogenesis and function. *Nature reviews genetics*, *17*(1), 47.

Ranzani, V., Rossetti, G., Panzeri, I., Arrigoni, A., Bonnal, R. J., Curti, S., . . . & De Francesco, R. (2015). The long intergenic noncoding RNA landscape of human lymphocytes highlights the regulation of T cell differentiation by linc-MAF-4. *Nature immunology*, *16*(3), 318.

Rapicavoli, N. A., Qu, K., Zhang, J., Mikhail, M., Laberge, R. M., & Chang, H. Y. (2013). A mammalian pseudogene lncRNA at the interface of inflammation and anti-inflammatory therapeutics. *Elife*, *2*.

Rice, K. L., & Licht, J. D. (2007). HOX deregulation in acute myeloid leukemia. *Journal of clinical investigation*, *117*(4), 865–68.

Rossetto, C. C., & Pari, G. S. (2014). PAN's labyrinth: molecular biology of Kaposi's sarcoma-associated herpesvirus (KSHV) PAN RNA, a multifunctional long noncoding RNA. *Viruses*, *6*(11), 4212–26.

Rossetto, C. C., Tarrant-Elorza, M., Verma, S., Purushothaman, P., & Pari, G. S. (2013). Regulation of viral and cellular gene expression by Kaposi's sarcoma-associated herpesvirus polyadenylated nuclear RNA. *Journal of virology*, *87*(10), 5540–53.

Sehgal, L., Mathur, R., Braun, F. K., Wise, J. F., Berkova, Z., Neelapu, S., . . . & Samaniego, F. (2014). FAS-antisense 1 lncRNA and production of soluble versus membrane Fas in B-cell lymphoma. *Leukemia*, *28*(12), 2376.

Sharma, S., Findlay, G. M., Bandukwala, H. S., Oberdoerffer, S., Baust, B., Li, Z., . . . & Rao, A. (2011). Dephosphorylation of the nuclear factor of activated T cells (NFAT) transcription factor is regulated by an RNA-protein scaffold complex. *Proceedings of the National Academy of Sciences*, *108*(28), 11381–86.

Shinjyo, T., Kuribara, R., Inukai, T., Hosoi, H., Kinoshita, T., Miyajima, A., . . . & Inaba, T. (2001). Downregulation of Bim, a proapoptotic relative of Bcl-2, is a pivotal step in cytokine-initiated survival signaling in murine hematopoietic progenitors. *Molecular and cellular biology*, *21*(3), 854–64.

Smith, W. L., DeWitt, D. L., & Garavito, R. M. (2000). Cyclooxygenases: structural, cellular, and molecular biology. *Annual review of biochemistry*, *69*(1), 145–82.

Sonkoly, E., Ståhle, M., & Pivarcsi, A. (2008). MicroRNAs: novel regulators in skin inflammation. *Clinical and experimental dermatology*, *33*(3), 312–15.

Spurlock III, C. F., Tossberg, J. T., Guo, Y., Collier, S. P., Crooke III, P. S., & Aune, T. M. (2015). Expression and functions of long noncoding RNAs during human T helper cell differentiation. *Nature communications*, *6*, 6932.

Sunwoo, H., Dinger, M. E., Wilusz, J. E., Amaral, P. P., Mattick, J. S., & Spector, D. L. (2009). MEN ε/β nuclear-retained non-coding RNAs are up-regulated upon muscle differentiation and are essential components of paraspeckles. *Genome research*, *19*(3), 347–59.

Szabo, S. J., Kim, S. T., Costa, G. L., Zhang, X., Fathman, C. G., & Glimcher, L. H. (2000). A novel transcription factor, T-bet, directs Th1 lineage commitment. *Cell*, *100*(6), 655–69.

Tavanez, J. P., Quina, A. S., & Cunha, C. (2014). Virus and noncoding RNAs: stars in the host–virus interaction game. *Future virology*, *9*(12), 1077–87.

Tong, Q., Gong, A. Y., Zhang, X. T., Lin, C., Ma, S., Chen, J., . . . & Chen, X. M. (2015). LincRNA-Cox2 modulates TNF-α–induced transcription of Il12b gene in intestinal epithelial cells through regulation of Mi-2/NuRD-mediated epigenetic histone modifications. *FASEB Journal*, *30*(3), 1187–97.

Turner, M., Galloway, A., & Vigorito, E. (2014). Noncoding RNA and its associated proteins as regulatory elements of the immune system. *Nature immunology*, *15*(6), 484.

Vahedi, G., Takahashi, H., Nakayamada, S., Sun, H. W., Sartorelli, V., Kanno, Y., & O'Shea, J. J. (2012). STATs shape the active enhancer landscape of T cell populations. *Cell*, *151*(5), 981–93.

Venkatraman, A., He, X. C., Thorvaldsen, J. L., Sugimura, R., Perry, J. M., Tao, F., . . . & Peng, L. (2013). Maternal imprinting at the H19–Igf2 locus maintains adult haematopoietic stem cell quiescence. *Nature*, *500*(7462), 345.

Verma-Gaur, J., Torkamani, A., Schaffer, L., Head, S. R., Schork, N. J., & Feeney, A. J. (2012). Noncoding transcription within the Igh distal VH region at PAIR elements affects the 3D structure of the Igh locus in pro-B cells. *Proceedings of the National Academy of Sciences*, *109*(42), 17004–9.

Vigneau, S., Rohrlich, P. S., Brahic, M., & Bureau, J. F. (2003). Tmevpg1, a candidate gene for the control of Theiler's virus persistence, could be implicated in the regulation of gamma interferon. *Journal of virology*, *77*(10), 5632–38.

Wang, P., Xue, Y., Han, Y., Lin, L., Wu, C., Xu, S., . . . & Cao, X. (2014). The STAT3-binding long noncoding RNA lnc-DC controls human dendritic cell differentiation. *Science*, *344*(6181), 310–13.

Wang, Y., Zhong, H., Xie, X., Chen, C. Y., Huang, D., Shen, L., . . . & Zeng, G. (2015). Long noncoding RNA derived from CD244 signaling epigenetically controls CD8+ T-cell immune responses in tuberculosis infection. *Proceedings of the National Academy of Sciences*, *112*(29), E3883–E3892.

Wei, L., Vahedi, G., Sun, H. W., Watford, W. T., Takatori, H., Ramos, H. L., . . . & Peng, W. (2010). Discrete roles of STAT4 and STAT6 transcription factors in tuning epigenetic modifications and transcription during T helper cell differentiation. *Immunity*, *32*(6), 840–51.

Willingham, A. T., Orth, A. P., Batalov, S., Peters, E. C., Wen, B. G., Aza-Blanc, P., . . . & Schultz, P. G. (2005). A strategy for probing the function of noncoding RNAs finds a repressor of NFAT. *Science*, *309*(5740), 1570–73.

Yagi, R., Zhu, J., & Paul, W. E. (2011). An updated view on transcription factor GATA3-mediated regulation of Th1 and Th2 cell differentiation. *International immunology*, *23*(7), 415–20.

Yoon, J. H., Abdelmohsen, K., & Gorospe, M. (2013). Posttranscriptional gene regulation by long noncoding RNA. *Journal of molecular biology*, *425*(19), 3723–30.

Zhang, Q., Chen, C. Y., Yedavalli, V. S., & Jeang, K. T. (2013). NEAT1 long noncoding RNA and paraspeckle bodies modulate HIV-1 posttranscriptional expression. *MBio*, *4*(1), e00596–12.

Zhang, X., Lian, Z., Padden, C., Gerstein, M. B., Rozowsky, J., Snyder, M., . . . & Newburger, P. E. (2009). A myelopoiesis-associated regulatory intergenic noncoding RNA transcript within the human HOXA cluster. *Blood*, *113*(11), 2526–34.

Zhu, J., Yamane, H., & Paul, W. E. (2009). Differentiation of effector CD4 T cell populations. *Annual review of immunology*, *28*, 445–89.

CHAPTER 5

LNCRNAS AS IMPORTANT MEDIATORS FOR DIAGNOSIS, PROGNOSIS, AND POTENTIAL THERAPEUTICS IN LEUKEMIAS, HEPATOCELLULAR CARCINOMA, AND BREAST CANCER

AASHIQ HUSSAIN BHAT,*[1]
KHALID BASHIR DAR,[1]
AND SHOWKAT AHMAD GANIE[1]

[1] Department of Clinical Biochemistry, faculty of Biological Science, University of Kashmir, Srinagar-190006, India
*Corresponding author:
Dr. Aashiq Hussain Bhat,
Cancer Diagnostic and Research Center, Skims, Soura, Srinagar-190011
Tel: 8825008327; E-mail: aashiqbht8@gmail.com

Abstract

Earlier lncRNAs (long non-coding RNAs) were considered to have no biological activity. However, transcriptome analysis indicated a large number of dynamically expressed lncRNAs with a variety of biological functions. But now it has been revealed that lncRNAs are important regulatory components of development, cell proliferation, and homeostasis. LncRNAs have been shown to have vital functions as oncogenes or tumor suppressors and also act as chromatin remodeling factors. LncRNAs also play important cell functions in cell nuclei and cytoplasm, in regulating DNA bending, insulation, transcription, splicing, translation, and stability, and in subnuclear organization and protein localization. LncRNAs play a valuable function to correctly diagnose some cancers and also help in the prognosis of some important cancers. Diagnosis and prognosis of cancers with lncRNAs is preferential than procedures like biopsies, because of noninvasiveness and could be done routinely as being part of body fluids.

In liver cancers lncRNAs HOTAIR, H19 is elevated to quite a good extent. Another lncRNA, MALAT1, is helpful in tumor maintenance, invasiveness, and in metastasis and in highly increased experimentally induced murine HCC (hepatocellular carcinoma) in humans. LncRNAs from recent studies have also emerged as the best regulators in breast cancer development, progression, and metastatic spread. Among them, lncRNA BORG is reported to drive breast cancer metastasis and disease recurrence. LncRNAs have shown a potential role in hematopoiesis and leukemogenesis as indicated by recent research. In certain hematological malignancies, lncRNAs identify a patient population at risk: it may classify patients into aggressive or mild cancer groups and also predict the patient's response to therapy. This chapter addresses some important finding and functions of lncRNA from reported literature about leukemias, hepatocellular carcinoma, and breast cancer and also discusses their diagnosis, prognosis, and achievements as potential therapeutics.

Keywords: lncRNAs, cancer, leukemias, hepatocellular carcinoma, breast cancer, therapeutics

Introduction

The total number of genes estimated earlier, 35,000–10,0000, has been reduced to just 25,000 protein-coding regions, which accounts for just 1–2% of the total genome (Liang et al. 2000). This fact encouraged the researchers to find the alternative functional elements in the genome. Scientists found a new group of molecules, non-coding RNAs, which have diverse functions in different processes, including development, physiology, and disease. These non-coding regions were found in abundance in the human genome (Carninci et al. 2005). The classification of non-coding RNAs is mainly based on the length of RNA transcripts. On the basis of its length, lncRNAs are divided into two groups: lncRNA longer than 200 nucleotides long and small noncoding RNAs shorter than 200 nucleotides long. Non-coding RNAs have wide functions in biological processes, including the regulation of gene expression; they also bring about epigenetic changes at definite loci (Peschansky and Wahlestedt 2014). Non-coding RNAs have a wide range of applications, which increase day by day because of their role in different diseases and various physiological processes. Presently the estimated number of lncRNAs is of the order of 10,000 transcripts, some of which play a critical role in the regulatory circuits for gene expression. But plenty of lncRNAs remain uncharacterized, except those that have biological applications that are debated hotly (Lee 2012).

The classification of lncRNAs is based on the comparative position of coding genes; as such, they are categorized into sense or antisense, or intronic if the lncRNA overlaps with another transcript. If transcription occurs in the vicinity of or in the opposite direction to a neighboring coding transcript then it is bidirectional; it is intervening or intergenic (lncRNAs) if it occurs in between two genes (Jia et al. 2010). Further, promoter or enhancer regions produce some lncRNAs, which are called pRNAs (or PROMPTs) and eRNAs, respectively. The action mechanism of the lncRNA varies in cytoplasmic and nuclear compartments. LncRNAs predominantly affect different cellular processes, which include a spatial conformation of chromosomes, chromatin and DNA modifications, pre-mRNA splicing, RNA transcription, mRNA degradation, and mRNA translation (Batista and Chang 2013). Further, some small biologically active peptides are also formed from them.

Promising pieces of evidence have proved that lncRNAs play vital regulatory functions in the progression and pathogenesis of cancers, including hepatocellular carcinoma (HCC), breast cancer, and leukemias. Thus, determining the vital functions of lncRNAs in cancers would pave the way to understanding the molecular mechanisms of these cancers. This in turn could help develop novel therapeutic targets. In this chapter, we will summarize some important lncRNAs in HCC, breast cancer, and leukemias, and their functional roles in these cancers, and will explore clinical implications for their possible role as diagnostic and prognostic biomarkers and molecular therapeutic targets.

LncRNAs in leukemias

In recent years, leukemia has witnessed an increasing rising trend in its incidence rate; unfortunately, however, its pathogenesis is not clear. Ample studies have highlighted the significance of lncRNAs in the pathophysiology of leukemia. Some of the important lncRNAs involved in leukemias are discussed below.

Acute lymphoblastic leukemia (ALL) related lncRNAs

T-ALL-R-LncR1

With the help of deep sequencing and high-throughput technology, the whole transcriptome of the Jurkat leukemia T-cell line was determined. In addition, T-ALL-R-LncR1 is expressed in neoplastic T lymphocytes children's samples with ALL, which indicates that T-ALL-R-LncR1 may be linked with T-ALL. This provided a new option for the targeted therapy and early diagnosis of T-ALL. Knocking down T-ALL-R-LncR1 results in a complex called the Par-4/THAP1 protein complex. It also raises the activity of a caspase-3 enzyme and as such causes the higher expression of the pro-apoptotic Smac protein (L. Zhang, Xu, and Lu 2014; Trimarchi et al. 2014). Thus, it is concluded that the lower levels of T-ALL-R-LncR1 contribute to Par-4-mediated apoptosis and the blocking of cellular proliferation.

LUNAR1

Trimarchi and his colleagues utilized RNA-sequencing techniques to focus on a NOTCH-regulated oncogenic, lncRNA leukemia-induced non-coding activator RNA-1(LUNAR1), which displays a high correlation with IGF1R, a receptor earlier implicated in T-ALL. Inhibition of NOTCH was found to down-regulate LUNAR1 expression, while its expression showed an increasing trend in primary T-ALL. Moreover, LUNAR1 showed high expression in T-ALL cases with notch mutation. Attenuating LUNAR1 expression by utilizing RNAi resulted in the down-regulation of IGF1R expression and decreased IGF1 pathway activity (Trimarchi et al. 2014; Xuefei Wang et al. 2015). LUNAR1 works like an enhancer of lncRNAs by employing a mediator complex to the IGF1R promoter. This in turn enhances the expression of IGF1R mRNA, maintaining IGF1 signaling and finally promoting the growth of T-ALL cells (Xuefei Wang et al. 2015). In general, we conclude that LUNAR1 could serve as an important biomarker for clinical trials as well as therapeutic protocols.

Acute myeloid leukemia (AML)-related lncRNAs

MEG3

AS a tumor suppressor gene, MEG3 may directly or indirectly enhance the anti-cancer effect via p53. Research shows the involvement of the aberrant promoter methylation of MEG3 in AML patients. MEG3 hypermethylation was found to reduce the survival rate of AML patients significantly (Benetatos et al. 2010). A study by Lv and coworkers found lower levels of lncRNA MEG3 in AML patients compared with healthy individuals. Research provides evidence of the implication of lncRNA MEG3 in the regulation of cell proliferation, apoptosis, and cell cycles. Moreover, studies also show that MEG3 may affect the growth of AML cells via the blocking of DNMT3A activity (Benetatos et al. 2010). Therefore, lncRNA MEG3 has prominent implications in the pathology of AML and could be targeted for therapeutic outcome.

RUNXOR

RUNXOR lncRNA is located inside the nucleus and its promoter is positioned 3.8 kb upstream of the RUNX1 promoter. RUNXOR shows an up-regulated trend in the bone marrow samples of AML patients. Moreover, the RUNXOR also exhibited increased expression in leukemia cells including K562, KG-1, KG-1α, and Kasumi-1. RUNXOR combines with the proximal and distal promoter and enhancers of RUNX1 through its 3-terminal fragment and also takes part in the orchestration of an intrachromosomal looping process. According to Wang and colleagues (P. Wang, Ren, and Sun 2012), RUNXOR uses its 3-terminal fragment to interact with ETO in chromosome 8, EVI1 in chromosome 3, and the fragile region of RUNX1 in chromosome 21. The above-mentioned process is believed to be involved in chromosome translocation by altering the spatial structure of the chromosome.

NEAT1

This lncRNA, named the nuclear enriched abundant transcript (NEAT), includes NEAT1-1 and NEAT1-2 (Naganuma and Hirose 2013). The expression of NEAT 1 decreases in APL cells (NB4 cell) and it has been found that promyelocytic leukemia-retinoic acid receptor α (PMLRARα) is an efficient repressor of the transcriptional process in APL cells (NB4 cell) (Zeng et al. 2014). Furthermore, it has been established that levels of NEAT1 are higher in ATRA-treated PL cells. All these pieces of evidence indicate the prominent role of NEAT1 in APL pathogenesis.

LLEST

LLEST shows down-regulation in myelodysplastic syndrome (MDS) and up-regulation in chronic aplastic anemia (CAA). Also, LLEST shows very low or even no expression in the marrow mononuclear cells of AML patients. Higher expression of LLEST in AML-M2 patients lowers the recurrence rate and also improves the prognosis. Moreover, LLEST is also implicated in inhibiting tumor growth via blocking the G2/M phase of the cell cycle and promoting cell apoptosis and differentiation (Qi et al. 2008; Sun et al. 2007). It was observed that the decrease of BCL-2 and the increased expression of caspase 9 were the major promoting factors of

LLEST-mediated apoptosis. Furthermore, scientists observed that the reduced levels of ATPase H+ transporting V1 subunit C1 (ATP6v1c1) decreased the survival of tumor cells.

UCA1

UCA1 (urothelial carcinoma-associated 1), an oncofoetal gene is implicated in carcinogenesis and embryonic development (Fan et al. 2014). In the case of myeloid differentiation, the CCAAT/enhancer-binding protein-α (C/EBPα) is an important component because its inactivation could lead to the development of AML. Employing AML cell lines to generate either C/EBPα-p30wild-type C/EBPα or wild-type C/EBPα, Hughes and co-workers found that the C/EBPα-p30 contributed to UCA1 transcription while the wild-type C/EBPα repressed the UCA1 expression. In addition, scientists also reported marked up-regulation of UCA1 in AML cases with CEBPA mutations. A study also reported an increased proliferation of AML cells due to the inhibition of the cell cycle regulator p27kip1 (Hughes et al. 2015). Thus, the studies clearly established that UCA1 can be a potential therapeutic target and a good diagnostic marker for AML as well.

Chronic myeloid leukemia (cml) related LncRNAs

LncRNA-BGL3

LncRNA Beta-Globin Locus 3 (LncRNA-BGL3) is the vital regulator for Bcr-Abl-mediated cellular transformation, which was identified in a Bcr-Abl-transformed human K562 leukemia cell line using an lncRNA cDNA microarray (G. Guo et al. 2015). Gene fusion between c-Ab1 (located on chromosome 9) and Bcr (located on chromosome 22) leads to the formation of Bcr-Ab1 and is involved after the cleavage and translocation. Oncoprotein (BCR-Abl) is a crucial factor that is needed for the inception, maintenance, and advancement of CML (Bono et al. 2016). It has been reported that Bcr-Abl decreases the levels of lncRNA-BGL3 in leukemia cells and in K562 cells, which are obtained from CML patients, mediated by the c-Myc-dependent methylation of DNA; on the other hand, its over-expression leads to apoptosis in Abl-transformed cells. Studies show that BGL3 lncRNA acts as ceRNA (competitive endogenous RNA) for binding miR-17, miR-106a, miR-93, miR-20a, miR-106b, and miR-20b to regulate expression of PTEN (G. Guo et al. 2015). More studies are needed in the future to evaluate lncRNA-BGL3 as a potential therapeutic agent in CML.

H19

Bcr-Abl-positive cell lines and primary CML cells express high levels of lncRNA H19 (2.3kb) obtained from patients. H19 silencing of leukemic cells, sensitized to undergo Imatinib-induced apoptosis, shows less Bcr-Abl tumor growth *in vivo*, which suggests a definite role for H19 in Bcr-Abl facilitated cell transformation. Furthermore, Guo et al. demonstrated that the silencing of Bcr-Abl decreases H19 expression. These findings reveal that H19 has an essential role in Bcr-Abl tumorigenesis; it has provided various beneficial insights into the mechanism of Bcr-Abl leukemogenesis. Despite these promising findings, a decrease in the expression of Bcr-Abl or kinase activity of Abl causes a reduction in c-Myc levels and as such reduces H19 levels simultaneously (Guijie Guo et al. 2014). Altogether, these studies provide evidence to support that the Bcr-Abl/c-Myc/H19 pathway plays a very important role in leukemogenesis.

MEG3

MEG3 is implicated not only in AML but also in CML. Growing evidence supports the theory that imatinib resistance has turned out to be a major hurdle for CML treatment; thus, it is imperative at the present time to look for a breakthrough (Salizzato et al. 2016). Zhou and his colleagues observed MEG3 expression in imatinib-resistant K562 cells and imatinib-resistant CML patients. However, an increase in MEG3 expression can reverse imatinib resistance and as such helps in cell apoptosis and cell-proliferation inhibition. Moreover, MEG3 negatively associates with miR-21, and via its regulation on miR-21, MEG3 is believed to contribute to imatinib resistance, and succeeding cell apoptosis/proliferation besides the expression of multidrug-resistant transporters such as ABCG2, MRP1, and MDR1 (X. Zhou et al. 2017).

LncRNAs in hepatocellular carcinomas (HCC)

Hepatocellular carcinoma is the major cancer of the liver. It is among the most prevalent cancers and is considered the third main cause of deaths (Altekruse, McGlynn, and Reichman 2009). Developing countries like Asian and African countries, which have a higher prevalence of HBV (hepatitis B virus) and HCV (hepatitis C virus), have higher mortality incidence of HCC because it is related to both these conditions (Varnholt et al. 2008). The poor prognosis for HCC is associated with disease conditions like HCV, HBV, cirrhosis, alcohol-induced liver diseases, and diabetes, which are among several conditions. Conditions like cirrhosis and chronic liver disease predispose a person to hepatocellular carcinoma. Other factors that pose the risk are endocrine factors, nutritional factors, certain hereditary conditions like hemochromatosis, obesity, and environmental factors. Additionally, other causal factors that cause the onset of HCC are alcohol abuse, chronic HBV/HCV, and non-alcoholic steatohepatitis (Kirchner et al. 2010). The pathology of HCC is quite complex, which makes it challenging to find the main causative agents for its onset and development. The progress and advancement of diseases are associated with several protein-coding genes. These include genes involved in cell-cycle regulation, DNA damage-response genes, cell signaling, and apoptosis. But the irregular expression profile of these genes makes it difficult to create a model for the genes responsible for HCC (Racken and Carl-McGrath 2001).

Different treatment strategies with little success have been developed in the past due to limited knowledge and the non-uniformity of HCC symptoms. Long non-coding RNAs find a significant place in HCC development and are important regulators for HCC as well. LncRNAs have been developed as major regulators of genome dynamics. That in part has been achieved due to their highly specialized expression patterns in response to signals in different cell types (Dinger et al. 2008) being highly regulated in a spatiotemporal manner (K. C. Wang and Chang 2011). Although the functions of the lncRNAs have not been revealed fully, abundant lncRNAs are still functionally uncharacterized (Goff and Rinn 2015). The possible role of lncRNAs in HCC within different stages is an exciting and rapidly evolving area of research. This particular section of the chapter deals with some significant studies and highlights the roles of various lncRNAs in HCC. Chromosomal instability in HCC is also attributed to the shortening and dysfunction of telomeres. Additional studies from different countries have reported a loss of alleles from chromosome arms, protein-coding gene mutations like oncogene β catenin (CTNNB1) and Janus kinase 1 (JAK1), the (p53) tumor suppressor gene, telomerase reverse transcriptase (TERT), (CDKN2A) cyclin-dependent kinase inhibitor 2A, fibroblast growth factor 19 (FGF19), myeloid-lymphoid leukemia (MLL), and AXIN1, and that phosphatase and tensin homolog (PTEN) are associated with patients having HCC (Kawai-Kitahata et al. 2016). It is because of this JAK/STAT and Wnt/β-catenin that cellular signaling pathways are considered most affected in HCC. The genes involved in these pathways possess mutations that are shown to be elevated with infections by HCV and HBV (Kawai-Kitahata et al. 2016; Tornesello et al. 2013). The advanced molecular techniques like exome sequencing revealed a group of genes highly affected in HCC, including the leucine rich repeat-containing family, chromatin remodelers, nucleotide-binding domain, a histone methyltransferase, calcium channel subunits, and oxidative stress associated genes (Gu et al. 2013). These pathways provided many molecular targets and some critical insights about the pathogenesis of HCC with potential targets for treatment, but there is a definite need to find more efficient and druggable targets. The present section provides useful insights about lncRNAs as potential therapeutic and prognostic targets in HCC.

HCC is among the most prevalent types of cancer and HCC cases are increasing at an alarming pace. The pathology of HCC is not well understood and the underlying causes are even less understood. Recently the bigger role of lncRNAs in HCC has been reported: various lncRNAs have been associated with HCC, including H19, MVIH, HULC, HEIH, MEG3, and TUC338. This section of the chapter discusses lncRNAs, their functions, and their role in the onset, progression, and apoptosis of liver cancer cells. A good number of lncRNAs are important regulators for the development of cancer, progression, and metastasis. Some important lncRNAs that are involved in HCC are discussed in detail in the sections below.

H19

H19 is expressed by a maternally imprinted gene and is 2.3 kb long. It acts as an oncogene as well as a tumor-suppressor gene. But the exact mechanism of action of H19 is still uncertain (Nordin et al. 2014). H19 is silent transcriptionally in adult tissue but is active during embryogenesis. It has been reported that this lncRNA is also expressed in the case of maternal under-nutrition. H19 lncRNA is involved in cancer pathogenesis, like angiogenesis and metastasis. H19 gene expression occurs in association with maternal genes, such as IGF-2 (insulin-like growth factor 2), which it is suggested has a role in cancer onset through various epigenetic modifications (K.-S. Kim and Lee 1997). This fact is further validated by the observed epigenetic abnormalities seen at IGF-2 and H19 loci in HCC (Thorgeirsson and Grisham 2002). But many studies have reported that a lot of different mechanisms are operating at the present locus so no single process can be attributed to cancer for onset and progression with certainty. H19's relationship for the development of HCC was later established by the knockdown of H19, which prevented HCC development (Matouk et al. 2007). One of the interesting characteristics of H19 lncRNA is the expression of miR-675 from its first exon, which as such acts as an miRNA reservoir as well as a sponge for miR-675 (Kallen et al. 2013; Raveh at al.

2015). Furthermore, H19 has also been reported to have versatile functions that include epigenetic ones.

HULC

HULC is an lncRNA that has 2 exons of 1.6 kb in length. HULC is up-regulated in HCC and colorectal cancers that metastasize the liver (Liu et al. 2012; Panzitt et al. 2007). The unregulated expression of HULC occurs with the transcription factor of the cyclic adenosine monophosphate response-element binding protein (CREB) in Hep3B cells (J. Wang et al. 2010). Also, in the case of HepG2, the HULC expression is also up-regulated with the interaction of HBV HBx protein with CREB (Du et al. 2012). HULC expression is regulated by another factor called insulin-like growth factor 2 mRNA binding protein (IGF2BP) (Hammerle et al. 2013). The higher expression of HULC is associated with a higher proliferation rate and tumor growth that is caused by the down-regulation of the tumor-suppressor gene p18. The expression of the transcription factor Peroxisome proliferator-activated receptor α (PPARA) is also influenced by HULC through the creation of methylation marks at the CpG island in the promoter of miR-9. Thus, it results in the increased expression of PPARA and that in turn elevates ACSL1 gene expression. This results in an increased recurrence of HCC rates (Cui et al. 2015). The proposed mechanism shows that HULC functions like a sponge for miR-372 and as such causes up-regulation of its target genes (J. Wang et al. 2010). Patients reported with HCC show increased levels of HULC; as such, HULC could be developed as a prognostic biomarker for HCC (Xie, Ma, and Zhou 2013).

MALAT1 (Metastasis-associated lung adenocarcinoma transcript 1)

During the cancerous state, chromosomal translocations and mutations are present in this gene locus (Shen et al. 2015). HCC cell lines also show the up-regulated expression of MALAT1 (Geng et al. 2011). Not only this, the differential expression of MALAT1 has been reported in various types of cancers like liver, lung, colon, prostate, ovaries, and breast. The expression of MALAT1 acts as a biomarker in the detection of the recurrence of liver cancer after a liver transplant; as such, it can be developed as a therapeutic target (Lai et al. 2012).

HOTAIR

HOTAIR is a 2.2kb-long lncRNA, transcribed from locus HOXC in an antisense manner; as such, it is called HOX antisense intergenic RNA. This lncRNA was first seen in fibroblast; it silences the HOXD locus by the trimethylation of H3K27.56, HOTAIR, by recruiting PRC2 and lysine-specific demethylase 1 (LSD1) protein complexes to the HOXD gene cluster, changing its histone methylation status and chromatin configuration. Hence, it results in tumor-suppressor gene silencing (Gupta et al. 2010; Tsai et al. 2010). This lncRNA is associated with breast, pancreatic, gastrointestinal, and colorectal cancers apart from HCC (Gupta et al. 2010; K. Kim et al. 2013). HOTAIR interacts to PRC2 by the 89-mer domain at the 5′ end sequence 110. This region is also called minHOTAIR as the protein LSD1 interacts with the 646-mer at the 3′ end. HOTAIR expression is increased in HCC, as is reported in the studies; its knockdown reduces hepatocellular carcinoma proliferation (Geng et al. 2011). Additionally, cells with HOTAIR lower expression render it more responsive to chemotherapy (Z. Yang et al. 2011).

DILC

Its expression was found to be less in HCC cell lines, as confirmed by microarray expression studies; as such, it was named as an lncRNA down-regulated in liver cancer (DILC). It was reported that this lncRNA inhibits the autocrine signaling pathway of IL-6/STST3 and thus suppress liver cancer (Xue Wang et al. 2016).

CCAT1 (colon cancer–associated transcript 1)

CCAT1 was first thought to be up-regulated in colon and gastric cancer (Nissan et al. 2012; Yang et al. 2013). It is present in the vicinity of the c-Myc gene and is 2.6 kb in length. The expression of this lncRNA was found to be higher in HCC than in noncancerous liver tissue. It gives a poor prognosis but aids the proliferation of HCC cancerous cells. CCAT1 controls the proliferation of cancer cells by working as a sponge for let-7 miRNA and as such raises the expression of c-Myc and HMGA2 (L. Deng et al. 2015). Thus, CCAT1 holds good therapeutic application as a molecular marker for the diagnosis of HCC and a target for HCC therapy.

HEIH

HEIH stands for high expression in HCC; it is an lncRNA whose expression is found to be high in HCC, indicated by its name as well. It is categorized as an oncogene and interacts with EZH2. Additionally, its high expression is related to the recurrence of hepatitis-B related HCC (HBV-HCC). This lncRNA has been used as a prognosis biomarker for HCC. Moreover, the increased expression of HEIH indicates the severity of HCC (F. Yang et al. 2011).

MEG3

MEG3 lncRNA is ~1.7kb in length and is expressed by a maternally imprinted gene (X. Zhang et al. 2009). MEG3 is observed to have down-regulation in HCC cell lines, and the reintroduction of this lncRNA decreases anchorage-dependent and anchorage-independent cell growth and apoptosis. Because of the absence of MEG3, studies have reported that the promoter regions of some genes show hypermethylation, which is involved in HCC. MEG3 down-regulated expression has been observed in a number of cancers, like lung, liver, cervical, and glioma (Lu et al. 2013; P. Wang et al. 2012; Y. Zhou, Zhang, and Klibanski 2012). For example, MEG3 levels were found to be reduced in cervical cancer tissues in contrast with healthy patients; furthermore, the levels correlated with an increase in tumor size and metastasis. Another study showed that MEG3 levels were most commonly down-regulated in HCC. This demonstrates its link with epigenetics, as 60% of cases reported in HCC were found with increased methylation. Another study reported that the progression of gastric cancer is regulated by MEG3 as it competes for the miR-181 family.

HOTTIP (HOXA transcript at the distal tip)

HOTTIP lncRNA is produced from the distal tip of gene HOXA13; its expression was found up-regulated in pancreatic and HCC cancers. The interaction between HOTTIP with a WD-repeat-containing protein 5/mixed lineage leukemia (WDR5/MLL) complex regulates the expression of HOXA locus. Under the regulation of the HOXA13 gene, it in turn creates a feedback regulation loop. A recent study has shown that miR-192 and miR-204 suppress the expression of HOTTIP with the involvement of Argonaute 2 (AGO2)-mediated RNA interference in HCC cells. HOTTIP suppression by these miRNAs reduces HCC cell viability and induced HOTTIP expression inhibits miRNAs' increased cellular proliferation.

With these findings, it became clear that HOTTIP and its related miRNAs could potentially act as a very good therapeutic strategy for HCC (Quagliata et al. 2014).

TUG1

TUG1 stands for taurine up-regulated gene 1 (TUG1); it is an lncRNA that is 7.1kb in length. This lncRNA is associated with bladder cancer, osteosarcoma, lung cancer, and liver cancer. It was first reported in retinal cells that respond to taurine signals in a screen designed to identify up-regulated genes. TUG1 expression is high in HCC. Cell growth is regulated by TUG1 as it causes epigenetic silencing of the kruppel-like factor 2 (KLF2) transcription factor. Additionally, the knockdown of TUG1 causes HCC cells to undergo apoptosis, as it does also in hepatoblastomas. With this background, for HCC treatment TUG1 has good therapeutic implications (Chen et al. 2015).

URHC

This is the commonly encountered lncRNA that has up-regulated expression in HCC, as is indicated by the name URHC (up-regulated in HCC). It was observed that URHC down-regulates ZAK and thereby assists in cell proliferation. This is because the down-regulation of URHC reduces the proliferation of cells and cause apoptosis in HCC cells. Thus, this lncRNA can be developed as a therapeutic in HCC therapy (Xu et al. 2014).

HCC and other lncRNA

Other lncRNAs are also associated with different aspects of HCC. Among them, the most validated includes microvascular invasion (MVIH) in HCCs, which is involved in HBV-related HCC tissues. Phosphoglycerate kinase 1 (PGK1) protein interacts with MVIH and results in increased tumor growth and angiogenesis (Yuan et al. 2012). Extra-coding CEBPA (ecCebpa) is another functional lncRNA that is involved in HCC and is obtained from the CEBPA gene. EcCebpa lncRNA binds the CEBPA genes promoter and also acts together with DNMT1; as such, it prevents the methylation of the promoter, which in turn leads to increased expression of CEBPA.211.

Thus, the above studies clearly show the involvement of lncRNA in HCC and also how it helped us understand the molecular mechanism operating in HCC. LncRNA regulates HCC development and progression;

thus, lncRNA serves diagnostic and therapeutic roles in HCC and other diseases.

LncRNAs in breast cancer

In females, breast cancer is the second most common cause of death. Diagnosis and management of this disease are quite complex, which predisposes breast cancers to shed malignant cells into circulation during the early developmental stages of the tumor, resulting in metastatic lesions that directly accounts for 90% of breast cancer deaths (Gooding et al. 2017). Knowledge of the proper genetic and epigenetic changes in the establishment of these deadly occult lesions is reasonably scarce and must be explored to frame therapeutic strategies that specifically inhibit the outgrowth of dispersed breast cancer cells. LncRNAs have the potent capacity to alter the proliferative, invasive, and metastatic capacity for malignant cells (Beckedorff et al. 2013), so oncogenic lncRNAs have a great potential to act as therapeutic targets and their inhibition have shown substantial efficacy for preclinical settings of metastatic diseases. In breast cancer, deregulation of lncRNAs like XIST, HOTAIR, H19, and MALAT has been confirmed in cell lines and samples of breast cancer. Furthermore, a strong association has been reported between the signature of the lncRNA and the survival rates of breast cancer patients. A study with next-generation sequencing showed the role of as many as 1300 lncRNAs with aberrant expression patterns in HER-2 subtypes of breast cancer. Moreover, studies reported more than 1750 lncRNAs to possess differential expression in TNBC (triple negative breast cancer) (Li and Luo 2016).

Thus, the studies clearly indicate that the aberrant expression pattern of lncRNAs may play a very critical part in the carcinogenesis of breast cancer. Furthermore, the expression of lncRNA in cells and tissues during the developmental stages is more specific than mRNA. Therefore, the pattern of expression of lncRNAs could be potentially used as a specific tool to distinguish different breast cancer subtypes. LncRNAs may also act as diagnostic biomarkers and potential therapeutic targets for individual diseases. Some important lncRNAs with aberrant expression are discussed below in detail.

H19

H19 over-expression in breast cancer promotes tumor progression and increases cell proliferation by promoting the G-S transition with the positive regulation of transcription factor E2F1 (Berteaux et al. 2005). An increase in H19 gene expression is observed in breast cancer as compared with healthy tissues. The higher expression of this gene is linked with tumor grades and the absence/presence of estrogen and progesterone receptors. Vennin and colleagues reported that the higher expression of miR-675/H19 increases the aggressiveness of breast cancer cells. It also increases cell proliferation and migration process *in vitro*; in addition, it enhances metastasis and tumor growth under *in vivo* conditions. Furthermore, it was identified that the CbI-b and c-CbI family (ubiquitin ligase E3) in breast cancer cells targets miR-675, which provided some new mechanistic insights about lncRNA H19 function in the advancement of breast cancer (Vennin et al. 2015).

PANDAR (promoter of CDKN1A antisense DNA damage activated RNA)

PANDAR is situated 5kb upstream of the CDKN1A (Cyclin-Dependent Kinase Inhibitor 1A) gene transcription start site. It gives rise to a 1.5kb lncRNA. Different controversial studies proved that PANDAR expression is increased in several cancers, such as gastric cancer, hepatocellular carcinoma, and breast cancer. This shows that it is a complex part in cancer. In the cell lines of breast cancer, it promotes tumor formation by G1/S transition regulation. PANDAR lncRNA-mediated cell growth promotion and G1/S transition occurs by the down-regulation of its downstream target, p16INK4A (Sang et al. 2016).

XIST (X inactive-specific transcript)

XIST is comprised of 17kb-spliced plus polyadenylated lncRNA and is encoded by a XIST gene situated on the X-chromosome in the X inactivation center. XIST is the main player in silencing the X-chromosome in female cells. All female somatic cells express this gene, but the expression is lost in ovarian and cervical cell lines and also in breast cancer cells. Normally, human cells have one X-chromosome active. But in cancers, there is a loss of the normal X chromosome with deregulated expression of XIST. Cancer cells show either up-regulation or down-regulation of XIST, signifying its controversial and complex function in cancer biology (Sirchia et al. 2009).

HOTAIR (HOX antisense intergenic RNA)

HOTAIR is present on the HOXC locus on chromosome 12q13.13. HOTAIR indirectly reduces the expression of miR-7 through HoxD10 inhibition in breast cancer and as such causes EMT progression. Micro RNA (miR-7) denotes a metastasis suppressing miRNA that can cause inverse EMT by down-regulating the STAT3 (signal transducer and activator of transcription 3) pathway in breast cancer cells (H. Zhang et al. 2014).

MALAT-1 (Metastasis-associated lung adenocarcinoma transcript 1)

MALAT-1 is situated on the chromosome 11q13 and codes for the 8kb lncRNA. As has been reported by recent studies, MALAT-1 is down-regulated in breast-tumor cell lines and in breast cancer tissue as well. MALAT-1 controls metastasis in breast cancer cells by inducing EMT through the activation of the PI3K-AKT (phosphatidylinositide 3-kinase-protein kinase B) pathway. In TNBC and HER2+ breast cancer, the expression of MALAT1 can prove a good prognostic marker, excluding luminal patients. In TNBC and HER2+ subtypes, levels of MALAT-1 predict metastasis and tumor recurrence in lymph-node-negative patients (Jadaliha et al. 2016).

GAS5 (Growth arrest-specific 5)

The GAS5 gene is situated on chromosome 1q25.1 and it gives rise to small nucleolar RNAs and PIWI-interacting micro RNAs as well as lncRNA. The first report of GAS5 shows its involvement in breast cancer. Mourtada-Maarabouni with his colleagues in 2009 reported the down-regulation of GAS5 in breast cancer tissues. GAS5 is commonly found down-regulated in different cancers with less expression and typically gives a poor prognosis (Mourtada-Maarabouni et al. 2009). GAS5 also promotes apoptosis and/or cell proliferation in various kinds of cells, including breast cancer cells. GAS5's role as a tumor suppressor is specified with its breast-tumor growth inhibition.

NBAT-1 (lncRNA neuroblastoma associated transcript-1)

The expression of NBAT-1 decreases in several cancers, including breast cancer, which suggests its possible role as a potential tumor suppressor. Low NBAT-1 expression levels in breast cancer are related to poor survival as well as the development of metastases in the lymph node. A study by Hu et al. (Hu et al. 2015) reported that NBAT-1 up-regulates DKK1 (Dickkopf WNT signaling pathway inhibitor 1) through the interaction with EZH2. DKK1 acts as an inhibitor for the WNT signaling pathway; as such, it suppresses the migration and invasion of breast cancer cells with the effect of NBAT-1.

DSCAM-AS1

DSCAM-AS1 is an antisense intronic lncRNA that depends on estrogen receptor α (ERα) and falls within the Down syndrome cell-adhesion molecule (DSCAM) gene. Particularly for the luminal breast cancer subtype, it is used as a biomarker. The knockdown of DSCAM-AS1 mimics several features, such as ERα silencing, apoptosis increase, the reduction of cellular-growth, and the initiation of EMT markers with no influence on ERα expression. In order to explore the DSCAM-AS1 functional mechanism further in-depth studies are required (Hajjari and Salavaty 2015).

Z38

Z38 is a newly discovered lncRNA that shows greater expression in breast cancer cells. The knockout of Z38 utilizing siRNAs suppresses tumorigenesis, cell proliferation, and tumor growth, and also causes apoptosis. Nevertheless, the complete functional mechanisms implicated as the regulatory functions of Z38 are still unclear in cancers (R. Deng et al. 2016).

Conclusion

LncRNAs are accepted as a significant part of the eukaryotic transcriptome. These lncRNAs play critical roles in diverse essential cellular processes, like chromatin remodeling, splicing, and epigenetic regulation and their dysregulation contributes to the advancement and progression of cancer. LncRNAs act through various mechanisms; their comprehensive characterization of the mechanism of action will help them to be used as potential diagnostic biomarkers and for the treatment of cancer patients in the future.

Conflict of interest

The authors declare that they have no conflicts of interest with respect to research, authorship, and publication of this book chapter.

Copyright and permission statement

We confirm that the materials included in this chapter do not violate copyright laws. Where relevant, appropriate permissions have been obtained from the original copyright holder(s). All original sources have been appropriately acknowledged and/or referenced.

Funding

No funding was received for this project.

References

Altekruse, S. F., McGlynn, K. A., & Reichman, M. E. (2009). Hepatocellular carcinoma incidence, mortality, and survival trends in the United States from 1975 to 2005. *Journal of clinical oncology*, *27*(9), 1485.

Batista, P. J., & Chang, H. Y. (2013). Long noncoding RNAs: cellular address codes in development and disease. *Cell*, *152*(6), 1298–307.

Beckedorff, F. C., Amaral, M. S., Deocesano-Pereira, C., & Verjovski-Almeida, S. (2013). *Long non-coding RNAs and their implications in cancer epigenetics*. Portland Press Limited.

Benetatos, L., Hatzimichael, E., Dasoula, A., Dranitsaris, G., Tsiara, S., Syrrou, M., . . . & Bourantas, K. L. (2010). CpG methylation analysis of the MEG3 and SNRPN imprinted genes in acute myeloid leukemia and myelodysplastic syndromes. *Leukemia research*, *34*(2), 148–53.

Berteaux, N., Lottin, S., Monté, D., Pinte, S., Quatannens, B., Coll, J., . . . & Adriaenssens, E. (2005). H19 mRNA-like noncoding RNA promotes breast cancer cell proliferation through positive control by E2F1. *Journal of biological chemistry*, *280*(33), 29625–36.

Bono, S., Lulli, M., D'Agostino, V. G., Di Gesualdo, F., Loffredo, R., Cipolleschi, M. G., . . . & Sbarba, P. D. (2016). Different BCR/Abl protein suppression patterns as a converging trait of chronic myeloid leukemia cell adaptation to energy restriction. *Oncotarget*, *7*(51), 84810.

Carninci, P., Kasukawa, T., Katayama, S., Gough, J., Frith, M. C., Maeda, N., . . . & Kodzius, R. (2005). The transcriptional landscape of the mammalian genome. *Science*, *309*(5740), 1559–63.

Huang, M. D., Chen, W. M., Qi, F. Z., Xia, R., Sun, M., Xu, T. P., . . . & Shu, Y. Q. (2015). Long non-coding RNA ANRIL is up-regulated in hepatocellular carcinoma and regulates cell proliferation by epigenetic silencing of KLF2. *Journal of hematology & oncology*, *8*(1), 57–57.

Cui, M., Xiao, Z., Wang, Y., Zheng, M., Song, T., Cai, X., . . . & Zhang, X. (2015). Long noncoding RNA HULC modulates abnormal lipid metabolism in Hepatoma cells through an miR-9–mediated RXRA signaling pathway. *Cancer research*, *75*(5), 846–57.

Deng, L., Yang, S.-B., Xu, F.-F., & Zhang, J.-H. (2015). Long noncoding RNA CCAT1 promotes hepatocellular carcinoma progression by functioning as let-7 sponge. *Journal of experimental & clinical cancer research, 34*(1), 18.

Deng, R., Liu, B., Wang, Y., Yan, F., Hu, S., Wang, H., . . . & Yang, Y. (2016). High expression of the newly found long noncoding RNA Z38 promotes cell proliferation and oncogenic activity in breast cancer. *Journal of cancer*, *7*(5), 576.

Dinger, M. E., Amaral, P. P., Mercer, T. R., Pang, K. C., Bruce, S. J., Gardiner, B. B., . . . & Sunkin, S. M. (2008). Long noncoding RNAs in mouse embryonic stem cell pluripotency and differentiation. *Genome research*, *18*(9), 1433–45.

Du, Y., Kong, G., You, X., Zhang, S., Zhang, T., Gao, Y., . . . & Zhang, X. (2012). Elevation of highly up-regulated in liver cancer (HULC) by hepatitis B virus X protein promotes hepatoma cell proliferation via down-regulating p18. *Journal of biological chemistry*, *287*(31), 26302–311.

Zhou, B., Zhuang, X. M., Wang, Y. Y., Lin, Z. Y., Zhang, D. M., Fan, S., . . . & Chen, W. L. (2017). Long non-coding RNA UCA1 increases chemoresistance of tongue cancer cells by regulating autophagy signaling. *International journal of oral and maxillofacial surgery*, *46*, 139.

Geng, Y. J., Xie, S. L., Li, Q., Ma, J., & Wang, G. Y. (2011). Large intervening non-coding RNA HOTAIR is associated with hepatocellular carcinoma progression. *Journal of international medical research*, *39*(6), 2119–28.

Goff, L. A., & Rinn, J. L. (2015). Linking RNA biology to lncRNAs. *Genome research*, *25*(10), 1456–65.

Gooding, A. J., Zhang, B., Jahanbani, F. K., Gilmore, H. L., Chang, J. C., & Valadkhan, S. (2017). The lncRNA BORG drives breast cancer metastasis and disease recurrence. *Scientific reports*, *7*(1), 12698.

Gu, D.-L., Chen, Y.-H., Shih, J.-H., Lin, C.-H., Jou, Y.-S., & Chen, C.-F. (2013). Target genes discovery through copy number alteration analysis in human hepatocellular carcinoma. *World journal of gastroenterology: WJG*, *19*(47), 8873.

Guo, G., Kang, Q., Chen, Q., Chen, Z., Wang, J., & Tan, L. (2014). High expression of long non-coding RNA H19 is required for efficient tumorigenesis induced by Bcr-Abl oncogene. *FEBS letters*, *588*(9), 1780–86.

Guo, G., Kang, Q., Zhu, X., Chen, Q., Wang, X., Chen, Y., . . . & Huang, S. (2015). A long noncoding RNA critically regulates Bcr-Abl-mediated cellular transformation by acting as a competitive endogenous RNA. *Oncogene*, *34*(14), 1768–79.

Gupta, R. A., Shah, N., Wang, K. C., Kim, J., Horlings, H. M., Wong, D. J., . . . & Wang, Y. (2010). Long non-coding RNA HOTAIR reprograms chromatin state to promote cancer metastasis. *Nature*, *464*(7291), 1071–76.

Hämmerle, M., Gutschner, T., Uckelmann, H., Ozgur, S., Fiskin, E., Gross, M., . . . & Schirmacher, P. (2013). Posttranscriptional destabilization of the liver-specific long noncoding RNA HULC by the IGF2 mRNA-binding protein 1 (IGF2BP1). *Hepatology*, *58*(5), 1703–12.

Hajjari, M., & Salavaty, A. (2015). HOTAIR: an oncogenic long non-coding RNA in different cancers. *Cancer biology & medicine*, *12*(1), 1.

Hu, P., Chu, J., Wu, Y., Sun, L., Lv, X., Zhu, Y., . . . & Su, S. (2015). NBAT1 suppresses breast cancer metastasis by regulating DKK1 via PRC2. *Oncotarget*, *6*(32), 32410.

Hughes, J. M., Legnini, I., Salvatori, B., Masciarelli, S., Marchioni, M., Fazi, F., . . . & Fatica, A. (2015). C/EBPα-p30 protein induces expression of the oncogenic long non-coding RNA UCA1 in acute myeloid leukemia. *Oncotarget*, *6*(21), 18534.

Jadaliha, M., Zong, X., Malakar, P., Ray, T., Singh, D. K., Freier, S. M., . . . & Prasanth, K. V. (2016). Functional and prognostic significance of long non-coding RNA MALAT1 as a metastasis driver in ER negative lymph node negative breast cancer. *Oncotarget*, *7*(26), 40418.

Jia, H., Osak, M., Bogu, G. K., Stanton, L. W., Johnson, R., & Lipovich, L. (2010). Genome-wide computational identification and manual annotation of human long noncoding RNA genes. *Rna*, *16*(8), 1478–87.

Kallen, A. N., Zhou, X. B., Xu, J., Qiao, C., Ma, J., Yan, L., . . . & Min, W. (2013). The imprinted H19 lncRNA antagonizes let-7 microRNAs. *Molecular cell*, *52*(1), 101–12.

Kawai-Kitahata, F., Asahina, Y., Tanaka, S., Kakinuma, S., Murakawa, M., Nitta, S., . . . & Nagata, H. (2016). Comprehensive analyses of mutations and hepatitis B virus integration in hepatocellular carcinoma with clinicopathological features. *Journal of gastroenterology*, *51*(5), 473–86.

Kim, K.-S., & Lee, Y.-I. (1997). Biallelic expression of the H19 and IGF2 genes in hepatocellular carcinoma. *Cancer letters*, *119*(2), 143–48.

Kim, K., Jutooru, I., Chadalapaka, G., Johnson, G., Frank, J., Burghardt, R., . . . & Safe, S. (2013). HOTAIR is a negative prognostic factor and exhibits pro-oncogenic activity in pancreatic cancer. *Oncogene*, *32*(13), 1616–25.

Kirchner, G., Kirovski, G., Hebestreit, A., Schölmerich, J., Schlitt, H. J., Stoeltzing, O., & Hellerbrand, C. (2010). Epidemiology and survival of patients with hepatocellular carcinoma in Southern Germany. *International journal of clinical and experimental medicine*, *3*(2), 169.

Lai, M. C., Yang, Z., Zhou, L., Zhu, Q. Q., Xie, H. Y., Zhang, F., . . . & Zheng, S. S. (2012). Long non-coding RNA MALAT-1 overexpression predicts tumor recurrence of hepatocellular carcinoma after liver transplantation. *Medical oncology*, *29*(3), 1810–16.

Lee, J. T. (2012). Epigenetic regulation by long noncoding RNAs. *Science*, *338*(6113), 1435–39.

Li, R., & Luo, Y. (2016). lncRNA and its biological function. *Journal of agricultural biotechnology*, *24*(4), 600–612.

Liang, F., Holt, I., Pertea, G., Karamycheva, S., Salzberg, S. L., & Quackenbush, J. (2000). Gene index analysis of the human genome estimates approximately 120,000 genes. *Nature genetics*, *25*(2), 239.

Liu, Y., Pan, S., Liu, L., Zhai, X., Liu, J., & Wen, J. (2012). A genetic variant in long non-coding RNA HULC contributes to risk of HBV-related hepatocellular carcinoma in a Chinese population. *PloS one*, *7*(4), e35145.

Lu, K. H., Li, W., Liu, X. H., Sun, M., Zhang, M. L., Wu, W. Q., . . . & Hou, Y. Y. (2013). Long non-coding RNA MEG3 inhibits NSCLC cells proliferation and induces apoptosis by affecting p53 expression. *BMC cancer*, *13*(1), 461.

Matouk, I. J., DeGroot, N., Mezan, S., Ayesh, S., Abu-lail, R., & Hochberg, A. (2007). The H19 non-coding RNA is essential for human tumor growth. *PloS one*, *2*(9), e845.

Mourtada-Maarabouni, M., Pickard, M. R., Hedge, V. L., Farzaneh, F., & Williams, G. T. (2009). GAS5, a non-protein-coding RNA, controls apoptosis and is downregulated in breast cancer. *Oncogene*, *28*(2), 195.

Naganuma, T., & Hirose, T. (2013). Paraspeckle formation during the biogenesis of long non-coding RNAs. *RNA biology*, *10*(3), 456–61.

Nissan, A., Stojadinovic, A., Mitrani-Rosenbaum, S., Halle, D., Grinbaum, R., Roistacher, M., . . . & Bostanci, E. B. (2012). Colon cancer associated transcript-1: A novel RNA expressed in malignant and pre-malignant human tissues. *International journal of cancer*, *130*(7), 1598–06.

Nordin, M., Bergman, D., Halje, M., Engström, W. & Ward, A. (2014). Epigenetic regulation of the Igf2/H19 gene cluster. *Cell proliferation*, *47*(3), 189–99.

Panzitt, K., Tschernatsch, M. M. O., Guelly, C., Moustafa, T., Stradner, M., & Strohmaier, H. M. (2007). Characterization of HULC, a novel gene with striking up-regulation in hepatocellular carcinoma, as noncoding RNA. *Gastroenterology*, *132*(1), 330–42.

Peschansky, V. J., & Wahlestedt, C. (2014). Non-coding RNAs as direct and indirect modulators of epigenetic regulation. *Epigenetics*, *9*(1), 3–12.

Qi, X., Chen, Z., Liu, D., Cen, J., & Gu, M. (2008). Expression of Dlk1 gene in myelodysplastic syndrome determined by microarray, and its effects on leukemia cells. *International journal of molecular medicine*, *22*(1), 61–68.

Quagliata, L., Matter, M. S., Piscuoglio, S., Arabi, L., Ruiz, C., Procino, A., . . . & Andersen, J. B. (2014). Long noncoding RNA HOTTIP/HOXA13 expression is associated with disease progression and predicts outcome in hepatocellular carcinoma patients. *Hepatology*, *59*(3), 911–23.

Racken, C., & Carl-McGrath, S. (2001). Pathology and pathogenesis of hepatocellular carcinoma. *Digestive diseases*, *19*(4), 269–78.

Raveh, E., Matouk, I.J., Gilon, M. & Hochberg, A. (2015). The H19 Long non-coding RNA in cancer initiation, progression and metastasis–a proposed unifying theory. *Molecular cancer*, *14*(1), 184.

Salizzato, V., Borgo, C., Cesaro, L., Pinna, L. A., & Donella-Deana, A. (2016). Inhibition of protein kinase CK2 by CX-5011 counteracts imatinib-resistance preventing rpS6 phosphorylation in chronic myeloid leukaemia cells: new combined therapeutic strategies. *Oncotarget*, *7*(14), 18204.

Sang, Y., Tang, J., Li, S., Li, L., Tang, X., & Cheng, C. (2016). LncRNA PANDAR regulates the G1/S transition of breast cancer cells by suppressing p16 INK4A expression. *Scientific reports*, *6*, 22366.

Shen, L., Chen, L., Wang, Y., Jiang, X., Xia, H., & Zhuang, Z. (2015). Long noncoding RNA MALAT1 promotes brain metastasis by inducing epithelial-mesenchymal transition in lung cancer. *Journal of neuro-oncology*, *121*(1), 101–8.

Sirchia, S. M., Tabano, S., Monti, L., Recalcati, M. P., Gariboldi, M., & Grati, F. R. (2009). Misbehaviour of XIST RNA in breast cancer cells. *PloS one*, *4*(5), e5559.

Sun, L., Gao, J., Yuan, L. X., Chen, T. T., Pan, L. L., & Zhou, C. Y. (2007). Expression of mitochondrial ferritin in K562 leukemic cell during TPA-induced cell differentiation. *Zhongguo shi yan xue ye xue za zhi*, *15*(2), 272–77.

Thorgeirsson, S. S., & Grisham, J. W. (2002). Molecular pathogenesis of human hepatocellular carcinoma. *Nature genetics*, *31*(4), 339.

Tornesello, M. L., Buonaguro, L., Tatangelo, F., Botti, G., Izzo, F., & Buonaguro, F. M. (2013). Mutations in TP53, CTNNB1 and PIK3CA genes in hepatocellular carcinoma associated with hepatitis B and hepatitis C virus infections. *Genomics*, *102*(2), 74–83.

Trimarchi, T., Bilal, E., Ntziachristos, P., Fabbri, G., Dalla-Favera, R., & Tsirigos, A. (2014). Genome-wide mapping and characterization of Notch-regulated long noncoding RNAs in acute leukemia. *Cell*, *158*(3), 593–606.

Tsai, M.-C., Manor, O., Wan, Y., Mosammaparast, N., Wang, J. K., & Lan, F. (2010). Long noncoding RNA as modular scaffold of histone modification complexes. *Science*, *329*(5992), 689–93.

Varnholt, H., Drebber, U., Schulze, F., Wedemeyer, I., Schirmacher, P., Dienes, H.P. & Odenthal, M. (2008). MicroRNA gene expression profile of hepatitis C virus–associated hepatocellular carcinoma. *Hepatology*, *47*(4), 1223–32.

Vennin, C., Spruyt, N., Dahmani, F., Julien, S., Bertucci, F. o., Finetti, P. (2015). H19 non-coding RNA-derived miR-675 enhances tumorigenesis and metastasis of breast cancer cells by downregulating c-Cbl and Cbl-b. *Oncotarget*, *6*(30), 29209.

Wang, J., Liu, X., Wu, H., Ni, P., Gu, Z., & Qiao, Y. (2010). CREB up-regulates long non-coding RNA, HULC expression through interaction with microRNA-372 in liver cancer. *Nucleic acids research*, *38*(16), 5366–83.

Wang, K. C., & Chang, H. Y. (2011). Molecular mechanisms of long noncoding RNAs. *Molecular cell*, *43*(6), 904–14.

Wang, P., Ren, Z. & Sun, P. (2012). Overexpression of the long non-coding RNA MEG3 impairs *in vitro* glioma cell proliferation. *Journal of cellular biochemistry*, *113*(6), 1868–74.

Wang, X., Chen, K., Guo, G., & Chen, J.-L. (2015). Noncoding RNAs and their functional involvement in regulation of chronic myeloid leukemia. *Briefings in functional genomics*, *15*(3), 239–48.

Wang, X., Sun, W., Shen, W., Xia, M., Chen, C., & Xiang, D. (2016). Long non-coding RNA DILC regulates liver cancer stem cells via IL-6/STAT3 axis. *Journal of hepatology*, *64*(6), 1283–94.

Xie, H., Ma, H., & Zhou, D. (2013). Plasma HULC as a promising novel biomarker for the detection of hepatocellular carcinoma. *BioMed research international*, 2013.

Xu, W.-H., Zhang, J.-B., Dang, Z., Li, X., Zhou, T., & Liu, J. (2014). Long non-coding RNA URHC regulates cell proliferation and apoptosis via ZAK through the ERK/MAPK signaling pathway in hepatocellular carcinoma. *International journal of biological sciences*, *10*(7), 664.

Yang, F., Xue, X., Bi, J., Zheng, L., Zhi, K., & Gu, Y. (2013). Long noncoding RNA CCAT1, which could be activated by c-Myc, promotes the progression of gastric carcinoma. *Journal of cancer research and clinical oncology*, *139*(3), 437–45.

Yang, F., Zhang, L., Huo, X. s., Yuan, J. h., Xu, D., & Yuan, S. x. (2011). Long noncoding RNA high expression in hepatocellular carcinoma facilitates tumor growth through enhancer of zeste homolog 2 in humans. *Hepatology*, *54*(5), 1679–89.

Yang, Z., Zhou, L., Wu, L.-M., Lai, M.-C., Xie, H.-Y., & Zhang, F. (2011). Overexpression of long non-coding RNA HOTAIR predicts tumor recurrence in hepatocellular carcinoma patients following liver transplantation. *Annals of surgical oncology*, *18*(5), 1243–50.

Yuan, S. X., Yang, F., Yang, Y., Tao, Q. F., Zhang, J., Huang, G., . . . & Zhang, L. (2012). Long noncoding RNA associated with microvascular invasion in hepatocellular carcinoma promotes angiogenesis and serves as a predictor for hepatocellular carcinoma patients' poor recurrence-free survival after hepatectomy. *Hepatology*, *56*(6), 2231–41.

Zeng, C., Xu, Y., Xu, L., Yu, X., Cheng, J., & Yang, L. (2014). Inhibition of long non-coding RNA NEAT1 impairs myeloid differentiation in acute promyelocytic leukemia cells. *BMC cancer*, *14*(1), 693.

Zhang, H., Cai, K., Wang, J., Wang, X., Cheng, K., Shi, F., . . . & Dou, J. (2014). MiR-7, inhibited indirectly by lincRNA HOTAIR, directly inhibits SETDB1 and reverses the EMT of breast cancer stem cells by downregulating the STAT3 pathway. *Stem cells*, *32*(11), 2858–68.

Zhang, L., Xu, H.-G., & Lu, C. (2014). A novel long non-coding RNA T-ALL-R-LncR1 knockdown and Par-4 cooperate to induce cellular apoptosis in T-cell acute lymphoblastic leukemia cells. *Leukemia & lymphoma*, *55*(6), 1373–82.

Zhang, X., Rice, K., Wang, Y., Chen, W., Zhong, Y., & Nakayama, Y. (2009). Maternally expressed gene 3 (MEG3) noncoding ribonucleic acid: isoform structure, expression, and functions. *Endocrinology*, *151*(3), 939–47.

Zhou, X., Yuan, P., Liu, Q., & Liu, Z. (2017). LncRNA MEG3 regulates imatinib resistance in chronic myeloid leukemia via suppressing microRNA-21. *Biomolecules & therapeutics*, *25*(5), 490.

Zhou, Y., Zhang, X., & Klibanski, A. (2012). MEG3 noncoding RNA: a tumor suppressor. *Journal of molecular endocrinology*, *48*(3), R45–R53.

CHAPTER 6

LONG NON-CODING RNA IN NEUROLOGICAL DISEASES: A MECHANISTIC INSIGHT

KHALID BASHIR DAR,[1]
AASHIQ HUSSAIN BHAT,[1] SHAJRUL AMIN,[1]
AND SHOWKAT AHMAD GANIE*[1]

[1] Department of Clinical Biochemistry/Biochemistry, School of Biological Sciences, University of Kashmir, Srinagar-190006, India
*Corresponding author:
Dr. Showkat Ahmad Ganie, Department of Clinical Biochemistry, University of Kashmir India
Tel: 9419972678; E-mail: showkatganie@kashmiruniversity.ac.in

Abstract

The central nervous system (CNS) is a highly complex biological system composed of many cell types that work in a highly coordinated manner. The development and functioning of CNS depends upon the controlled expression of genes in its constituent neurons. The derailment of complex regulatory networks often leads to dreadful pathologies like Huntington's disease (HD), Alzheimer's disease (AD), and Parkinson's disease (PD). Neurodegenerative disorders occur not only due to the genes that code for proteins, but also due to various non-coding RNAs (ncRNAs). Long non-coding RNAs (lncRNAs) are an important class of ncRNAs considered to be prominent culprits for such diseases. At the least, 10,000 lncRNAs have been identified in the human genome so far. Although not understood comprehensively, lncRNAs are widely expressed in the nervous system of mammals and play a crucial role in the development, function, and disease pathology of CNS. In fact, lncRNAs are implicated in bringing about gene regulation at transcriptional, post-transcriptional, and epigenetic levels via a varied range of mechanisms. Interestingly, many lncRNAs show high expression in the adult and developing brain, often exhibiting precise temporal and spatial patterns of expression that indicate that they have imperative roles in CNS development, function, and disease pathology. LncRNAs like BACE1-AS, BC-RNA, GDNFOS, naPINK1, NEAT1, NEAT2, UBE3A-ATS, 51A, HOTAIR, and AS-Uchl1 have a prominent implication in the pathology of various neurodegenerative diseases and could serve as potential targets for treating such dreadful diseases. Understanding the active mechanisms of these crucial lncRNAs in neurodegenerative diseases would definitely provide new future insights into the possible interventions of these short oligonucleotides as therapeutic targets.

Keywords: non-coding RNA, neurodegeneration, amyotrophic lateral sclerosis, Angelman syndrome, Parkinson's disease, Alzheimer's disease

Introduction

Previously, 98% of human DNA was considered to be junk DNA in the context of being non-coding; however, it is now termed "non-coding DNA." This ncDNA transcribes to ncRNA, which includes tRNA, rRNA, and small RNAs such as microRNAs, siRNAs, piRNAs, snoRNAs, snRNAs, exRNAs, scaRNAs, and lncRNAs. The enormous amount of ncRNA inside the cell begs the question of its functionality. Technologies like microarrays and next-generation sequencing (deep sequencing) have shown that most of this ncDNA transcribes to ncRNA. Evidence such as (1) the conservation of promoter sequences, (2) expression patterns and responsiveness to stimuli that are specific to tissue/cell/sub-cellular structures, and (3) dynamic expression during development and disease strongly indicate the functionality of these ncRNAs (Ulitsky et al. 2011). Nevertheless, further studies are required to fully understand the role of each of these non-coding transcripts in molecular pathways. NcRNA shows interactions including RNA–RNA, RNA–DNA, and RNA–protein interactions that affect crucial processes like transcription, translation, alternative splicing, and epigenetic events, notably chromatin modification and RNA editing (Cech et al. 2014). Due to their broad biological actions, ncRNAs are recognized to be contributing factors in many diseases including neurodegenerative disorders. The human genome transcribes several thousand lncRNA transcripts with no potential to code for proteins but can regulate crucial biological functions. In fact, abnormal lncRNA expression may lead to significant neural abnormalities like schizophrenia, autism, and developmental delay, as well as Huntington's and Parkinson's diseases. Unfortunately, very few lncRNAs have been characterized in terms of their function and this lack of knowledge is a big hindrance in the progress of the emerging field of lncRNA biology. Studying individual lncRNAs is critical for understanding their role in disease pathology and to decode their potential as therapeutic targets (Cech et al. 2014). Before the advent of high-throughput sequencing (HTS) technology, very few lncRNAs were known. For instance, the lncRNA XIST was discovered in the early 1990s and later studies indicated its critical role in X-chromosome inactivation (Lessing et al. 2013). Since then, many lncRNAs have been discovered. Currently, there are over fifty thousand human lncRNAs (Iyer et al. 2015). At present, a meager amount of data is available regarding the role of lncRNAs in the molecular mechanics of the nervous system. To understand the biological role of lncRNAs, there is a need for the selection of specific lncRNAs, the determination of their roles in *in vivo* experiments, and exploring their mechanisms of action. As per the GENCODE Consortium, 40% of differentially expressed lncRNAs are specific to the brain. Even in the brain, lncRNAs show region-specific expression patterns and are expressed dynamically during neural differentiation (*in vitro*) and NSC populations (*in vivo*) (Ramos et al. 2013).

LncRNAs and their implications in CNS

Transcripts greater than 200 nucleotides (nt) are considered to be lncRNAs and include a diverse range of regulatory molecules such as non-coding RNA repeats, natural anti-sense transcripts (NATs) and lincRNAs. LncRNAs are classified as sense, antisense, intronic, or intergenic, being bidirectional with reference to nearby protein-coding genes. LncRNAs are transcribed from either strand of DNA and most lncRNAs are processed as typical mRNAs including 5′ capping, polyadenylation, and histone modifications (Knauss and Sun 2013). When compared with protein-coding genes, the splicing of lncRNAs is less efficient and is even absent in some cases (Tilgner et al. 2012). Most lncRNAs found inside a nucleus are predominantly associated with chromatin; however, some lncRNAs migrate to cytoplasm and perform important functions (Derrien et al. 2012). Previously, lncRNAs were considered just a transcriptional noise due to their lack of sequence conservation and the low fidelity of RNA polymerase II transcription initiation (Struhl 2007). This low sequence conservation of lncRNAs is believed to preserve the secondary structure of lncRNAs rather than the primary sequence; thus, it may allow lncRNAs to interact with the same protein complexes (Knauss and Sun 2013). LncRNAs have multiple characteristic features (figure 1) that support the meaningful role of lncRNAs in different physiological mechanisms.

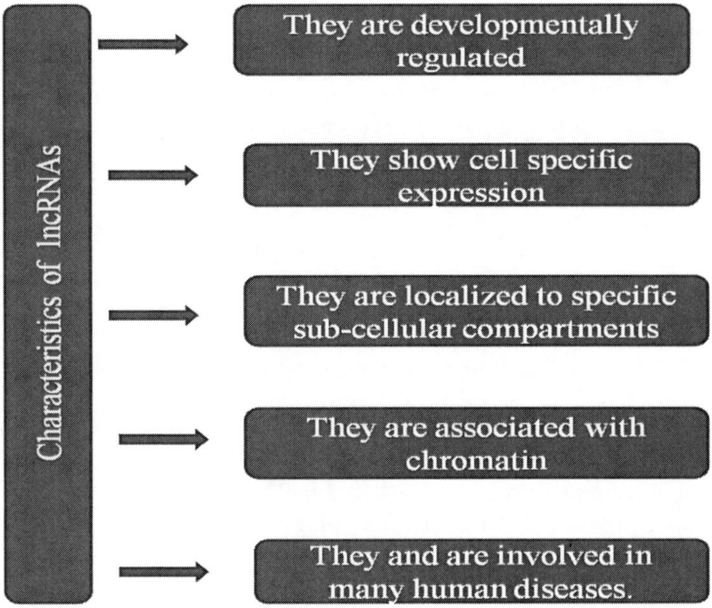

Figure 1. Characteristic features of lncRNAs.

The central nervous system (CNS) is a highly complex biological system consisting of regionally distinct neuronal and glial subtypes that are organized into dynamic neural networks responsible for various functions of the nervous system. There occurs abundant expression of lncRNAs in the cells of CNS and studies demonstrate that lncRNAs play critical roles in the spatial-temporal control of gene expression in brain development (Mercer et al. 2008). LncRNAs reside in discrete neuro-anatomical regions, specific cell types, or sub-cellular components, suggesting their role in biological functions. LncRNAs have a crucial role in the central nervous system, which implicates their connection to various neurological disorders (Ziats et al. 2013). Studies show a high conservation of lncRNAs between the brains of mice and humans, enriched in the predicted RNA secondary structure, preferentially located near protein-coding genes implicated in the development and function of the nervous system (Ponjavic et al. 2009). LncRNAs influence the pluripotency or differentiation in the cells of humans and mice, especially that of neural cells (Ng et al. 2012). Besides neuro-developmental disorders, lncRNAs are also implicated in neuro-oncological, neuro-immunological, and neurodegenerative disorders (Knauss and Sun 2013). During development, extracellular signals including growth factors, hormones, and neurotransmitters alter gene expression in a spatially and temporally coordinated way. In particular, extracellular stimuli, notably membrane depolarization and neuronal activity, play an important role in regulating the expression of genes in the central nervous system. The phenomenon of neuronal membrane depolarisation results in the entry of calcium through voltage-sensitive calcium ion channels. This influx of calcium triggers the calcium-dependent signaling pathways that ultimately have an effect on gene expression (Berridge 1998). In cultured neurons, many ncRNAs are induced due to neuronal activity (West et al. 2001). LncRNAs notably metastasis-associated lung adenocarcinoma transcript 1 (MALAT1) and nuclear-enriched abundant transcript 1 (NEAT1) are induced due to membrane depolarization (Lipovich et al. 2012). NEAT1 and MALAT1 are up-regulated during neuronal and glia differentiation (Mercer et al. 2010). Moreover, MALAT1 also regulates the process of synapse formation through the regulation of alternative splicing of synapse genes (Bernard et al. 2010). LncRNAs like NEAT1 and NEAT2 (also called MALAT1) when activated play an important role in modulating processes such as neuronal growth/branching and the development of g-aminobutyric acid (GABA) inter-neurons (Greer and Greenberg 2008; Wong et al. 2002; García et al. 2011) by regulating the expression of various genes. Brain-derived neuro-trophic factor antisense RNA (BDNF-AS) represents a prominent example of an activity-regulated lncRNA that influences gene expression in the brain. The BDNF gene encodes a secreted growth factor that enhances maturation and the survival of neurons.

The role of lncRNAs in neuronal differentiation

Since lncRNAs are expressed in a tissue-specific manner so most current research focuses on elucidating the mechanistic role of lncRNAs in the development of the nervous system. A series of developmental events occurs in the nervous system of an embryo and adult throughout life. Among these events, fate determination and the differentiation of neural stem cells are of prime importance due to their role in determining brain function. The systematic evaluation of differential expression profiles of lncRNAs during neuronal differentiation has led to the opinion that lncRNAs have definite roles in the determination of neural cell fate. Various lncRNAs are found involved in functions like neural lineage specification, neuronal-glia fate switching, and oligodendrocyte maturation (Mercer et al. 2010). The expression of lncRNAs such as Dlx1AS (distal-less homeobox 1- anti sense) lncRNA showed a down-regulated trend during oligodendrocyte differentiation and an up-regulated trend during GABAergic differentiation.

Dlx1AS has been previously reported to show co-expression with its overlapping gene, namely Dlx1 in the regions associated with neurogenesis (Dinger et al. 2008), thereby raising the possibility of its involvement in gene regulation during neurogenesis. Evf2 represents a good example of lncRNA that is transcribed both in human cells and mouse cells, especially during the development of the forebrain. It is important for the formation of GABAergic interneurons. Evf2 is transcribed from the intergenic region that lies between the Dlx5 and Dlx6 loci, and is overlapped by the conserved Dlx5/6 intergenic enhancer. In the case of developing telencephalon, Evf2 acts as a downstream target of sonic hedgehog (Shh) signaling (Feng et al. 2006). Also, in the developing ventral forebrain, Evf2 recruits transcription factors like Dlx and methyl CpG binding protein 2 (MeCP2) that regulate the expression of Dlx5, Dlx6, and glutamate de-carboxylase 1 (Gad1) through cis- and trans-acting mechanisms. Gad1 is responsible for catalyzing the synthesis of GABA from L-glutamic acid. Subsequently, mice deficient in Evf2 exhibit decreased numbers of GABAergic interneurons in the hippocampus and dentate gyrus (Bond et al. 2009). Since altered GABAergic interneuron function is implicated in neural disorders like epilepsy, autism, schizophrenia, and mental retardation, studies elucidating the mechanism through which Evf2 regulates gene expression in the GABAergic inter-neurons will prove highly beneficial for understanding the molecular basis of neuropsychiatric disorders. The Rhabdomyosarcoma 2–associated transcript (RMST) is another example of well-conserved, functional lncRNA having a critical role in neurogenesis. Recently, RMST was identified as an abundantly expressed lncRNA in dopaminergic neurons, which are essential for the differentiation of neurons (Ng et al. 2012). RMST is expressed highly in the midbrain–hindbrain boundary and explicitly in the dopaminergic neuronal precursors of the midbrain, where it is co-expressed with LIM homeobox transcription factor 1, alpha (Lmx1a), in the developing brain (Uhde et al. 2010). All such evidence shows that RMST has a prominent role in the differentiation of midbrain neurons. The enhancer region of neurogenin 1 (Neurog1) represents another example of an lncRNA modulating brain development. This enhancer RNA is transcribed actively in mice; and gives rise to an approximately 1.4kb nuclear lncRNA known as utNgn1. Later, it plays an important role in the positive regulation of Neurog1 expression in cortical development (Onoguchi et al. 2012). Knocking down utNgn1 in neo-cortical neural progenitors resulted in the inhibition of two important early markers of neocortical differentiation, namely neurogenic differentiation 1 (NeuroD1) and T-brain gene-2 (Tbr2) expression, thereby indicating that the utNgn1 transcript has a key role in cortical development. Thus there is growing evidence to support the importance of lncRNAs in the development of CNS, which implies that their deregulation can lead to various neurological disorders.

LncRNAs in Alzheimer's disease

Alzheimer's disease (AD), identified more than 100 years ago is a slowly progressive neurodegenerative disease characterized by memory loss and other cognitive dysfunction. It is the most common form of senile dementia. Most AD cases are late-onset, usually occurring after the age of 65 and accounting for about 90% of cases. The early-onset form of AD is more commonly inherited. So far, various lncRNAs have been implicated in the pathology of AD (figure 2); certain important lncRNAs are discussed below.

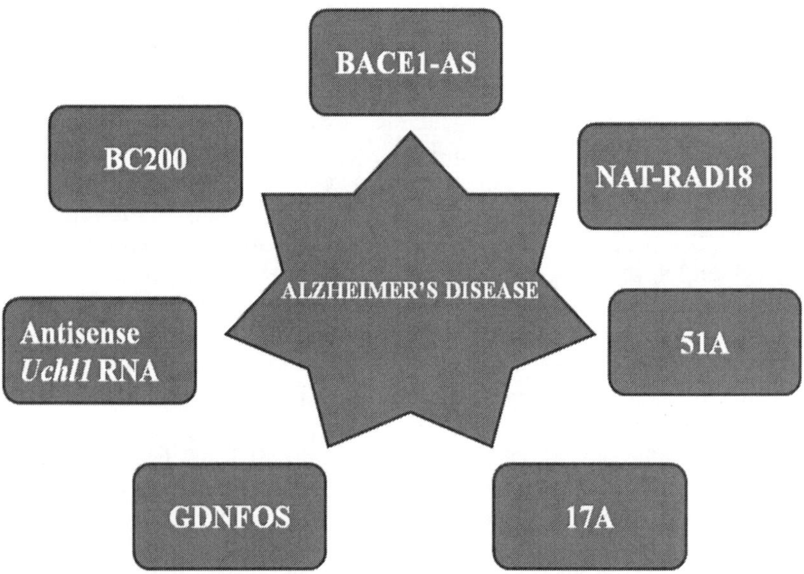

Figure 2. Various lncRNAs implicated in the pathology of Alzheimer's disease.

B-secretase-1 (BACE1)-AS lncRNA

β-Amyloid peptide (Aβ), the main component of senile plaque, is crucially involved in the pathological process of AD. There are two major isoforms of Aβ (Aβ42 and Aβ40) and they exist together in amyloid plaques. Aβ42 isoform has two extra residues at C-terminus and is considered the most amyloidogenic isoform. In the brains of certain AD patients, the levels of total Aβ levels as well as the ratios of Aβ42:Aβ40 are increased. Abundant evidence suggests that insoluble Aβ oligomers are more toxic than other peptide forms and correlate with the severity of dementia. When APP is first cleaved by the β-site APP-cleaving enzyme 1 (BACE1), it forms sAPPβ. The latter is handled with γ secretase to produce Aβ peptide, which accumulates in the neurons and causes synapse dysfunction, cognitive dysfunction, and the formation of plaques during AD. But when cleaved by α-secretase and γ-secretase, the final fragments have no toxic effect on the neurons in the brain. Therefore, BACE1 enzyme is critical for the production of beta-amyloids and the subsequent formation of amyloid plaques, which play a central role in the pathophysiology of AD. BACE1-AS has been found to upregulate the BACE1sense mRNA by duplexing with it and thereby stabilizing it, thus promoting the levels of BACE1 enzyme, which in turn increases levels of amyloid beta. MiR-485-5p is a well-known miRNA involved in BACE1 post-transcriptional regulation. It has been reported that BACE1-AS prevents MiR-485-5p-induced repression of BACE1 mRNA by masking the binding site. BACE1-AS can also bind with HuD, a primarily neuronal RNA-binding protein (implicated in learning and memory). This evidence indicates that BACE1-AS finely controls the expression of BACE1 through a variety of mechanisms. Now the concern is whether BACE1-AS is altered in the brain samples of AD patients. As per research-based evidence, the relative quantity of the BACE1-AS transcript in the parietal cortex, superior frontal gyrus, entorhinal cortex, and hippocampus of post-mortem brains increases along with a smaller increase in the expression of the BACE1 transcript. Cellular stressors such as hydrogen peroxide (H_2O_2), serum starvation, Aβ1-42, and high glucose have already been implicated in the pathogenesis of AD, and BACE1-AS has been found to respond to all these stresses. Therefore, increasing BACE1-AS caused by these factors may contribute to the development of AD by enhancing BACE1 expression. Overwhelming studies confirm elevated levels of BACE1-AS in the brains of AD patients, indicating that this lncRNA (BACE1-AS) is an important driving force behind BACE1 up-regulation (Modarresi et al. 2011). In order to understand the mechanism through which BACE1-AS is involved in the pathology of AD, scientists knocked down BACE1-AS by continuous infusion of locked nucleic acid (LNA)-modified small interfering RNAs (siRNAs) into mouse brains and down-regulated BACE1-AS, BACE1, and the amyloid beta formation/aggregation in the brain. Thus targeting BACE1-AS lncRNA could represent an effective approach for the development of effective anti-AD therapeutics.

Brain cytoplasmic (BC) lncRNA

BC RNAs represent the other important group of lncRNAs that are found deregulated in AD. For instance, BC200 RNA (in humans) and BC1 RNA (in mice) are transported as ribonucleoprotein molecules to dendritic processes and play an important role in modulating gene expression at the translational level (Muddashetty et al. 2002). Levels of BC200 were found to be higher in age-matched AD brains compared with normal brains, and there was a relative increase in the levels of BC200 RNA in affected areas of the brain with the severity of the AD (Mus et al. 2007). Studies demonstrate that BC200 RNA interacts with numerous RNA-binding proteins implicated in mRNA trafficking in neurons, for example (1) heterogeneous nuclear ribonucleoprotein A2 (hnRNPA2), which is involved in the transport of mRNA in neurons and oligodendrocytes (Muslimov et al. 2007); (2) poly(A)-binding protein (PABP1), the regulator of translation initiation (Muddashetty et al. 2002); and (3) synaptotagmin-binding cytoplasmic RNA interacting protein (SYNCRIP), which is part of the mRNA transport granule and is possibly implicated in local protein synthesis at postsynaptic sites in neuronal cells (Duning et al. 2008). Overall this evidence points out the credible function of BC200 lncRNA in modulating the synthesis of local proteins at the dendrites, and also indicates that over-expression of BC200 in AD brains could be one of the important causes of synapto-dendritic deterioration.

51A lncRNA

The neurons of the central and peripheral nervous system express the sortilin-related receptor 1 (Sorl1) gene, which is a susceptibility gene for late-onset Alzheimer's disease (Jin et al. 2013). Various variants of the Sorl1 gene are associated with AD-related brain morphologies (Assareh et al. 2014). In fact, there is reduced expression of Sorl1 in the brains of patients suffering from AD (Ma et al. 2009). During the process of APP translocation, Sorl1 interacts with APP and acts as the sorting receptor member protein. When over-expressed in neurons, Sorl1 causes the redistribution of APP and diminishes the transition of APP to endosomes. Loss of Sorl1 function affects the cleavage pathway of APP and also enhances the formation of neurotoxic Aβ peptides (Andersen et al. 2005). Therefore, Sorl1 restricts the processing of APP into Aβ and thus reduces the burden of amyloidogenic peptide formation. Thus, Sorl1 possesses neuroprotective functions. 51A is an antisense transcript of intron 1 of the SORL1 gene that alters the spliced form of SORL1 mRNA, thus resulting in Aβ42 accumulation (Guennewig and Cooper 2014). Because it is a culprit in the pathology of AD, it is obvious that the strategy to target 51A could play an imperative role in the treatment of this horrible disease.

17A lncRNA

17A is a novel lncRNA expressed in the human brain. It shows elevated levels in cerebral tissues of AD patients and is implicated in (1) the increased production of Aβ peptides, (2) GPR51 alternative splicing, and, finally, (3) impaired GABA B function. 17A is found embedded in an antisense orientation in the third intron of the human G-protein-coupled receptor 51 (GPR51) gene (also called GABBR2 gene), and has been implicated in regulating GPR51/GABBR2 pre-mRNA processing, and in promoting the production of the alternative and non-functional splicing isoform B of the GABA B receptor (GABAB R2). The stable expression of 17A in human neuroblastoma cells was found to promote the synthesis of GABAB R2 splicing isoform B. The latter is the defective form of the intra-membrane sequence peptide, thus generating non-functional receptors that are unable to transduce GABAB-dependent intracellular signaling. In general, the higher level of lncRNA 17A was found to promote the secretion of Aβ peptide with an increase in the Aβ42/Aβ40 ratio (Rege et al. 2013). The expression of 17A shows an increasing trend due to inflammatory stimuli, thus creating the possibility of using anti-inflammatory drugs with modulatory effects on Aβ synthesis. Down-regulating 17A lncRNA by using specific antisense oligonucleotide could represent a potential therapeutic approach for reducing Aβ production in AD; previously this approach had been found effective in AD animal and cell models. Moreover, the strategy to modulate lncRNAs to modify the subsequent expression of known target genes involved in neurodegeneration may represent another promising anti-AD therapeutic approach.

GDNFOS lncRNAs

GDNFs, a member of the nerve growth factor family of neurotrophins, have been implicated in promoting the survival of the dopamine-releasing neurons of the midbrain. In addition, the exogenous GDNF has also been found to promote the survival of other neurons including GABAergic and forebrain cholinergic neurons, peripheral sensory and autonomic neurons, and spinal motor neurons (Airaksinen and Saarma 2002). GDNF also performs functions like the regulation of neurite branching and synaptic plasticity, as well as promoting migration and differentiation of neural stem cells (Yoo et al. 2012). As far as the human GDNF gene is concerned, it has six exons. Transcription of five GDNF mRNAs occurs by alternative splicing. GDNF is synthesized as the precursor pre-pro- GDNF that is processed into mature/biologically active GDNF (Lin et al. 1993). Patients suffering from AD show significantly higher concentrations of mature GDNF in cerebrospinal fluid, and lower levels in serum (Straten et al. 2009). The middle temporal gyrus (MTG), which is consistently affected in AD patients, shows decreased levels of mature GDNF (Airavaara et al. 2011). Overall, this evidence suggests that there is a definite connection between the regulation of GDNF and the pathology of AD. GDNFOS, the cis-natural antisense transcript (NAT) transcribed from the opposite strand of the GDNF gene contains four exons and three initial sites to form three different splicing sequences, namely GDNFOS1, GDNFOS2, and GDNFOS3. GDNFOS overlaps with the GDNF 5-UTR exon. GDNFOS1 and GDNFOS2 are lncRNAs and GDNFOS3 contains a potential open-reading frame (ORF) that encodes a protein with no known homologue in GenBank. GDNFOS1 expression takes place in different areas of the human brain and its transcriptional level is lower compared with GDNF isoform containing exon 1 and exon 4. There is a belief that GDNFOS1 might function to regulate the higher levels of GDNF in the human brain (Zhang 2016). Future studies will focus more on understanding the interaction between GDNFOS and GDNF mRNA so as to explore its possible role in the mechanism of AD pathogenesis.

Other candidate lncRNAs in AD

Various lncRNAs including NATRAD18 and antisense Uchl1 RNA are speculated to have a possible role in the pathogenesis of Alzheimer's disease. NATRad18 lncRNA acts as a natural antisense transcript against Rad18. The later encodes a protein involved in repairing DNA. NAT-Rad18 is specifically expressed in neuronal cells and exhibits up-regulation in Aβ-treated rat cortical neurons (Parenti et al. 2007). On the other hand, antisense Uchl1 RNA is a nuclear-enriched lncRNA against mouse ubiquitin carboxy-terminal hydrolase L1 (Uchl1). This lncRNA controls synthesis of UCHL1 protein at post-transcriptional level (Carrieri et al. 2012). Despite UCHL1 participating in the progression of Alzheimer's disease (Zhang et al. 2014), until now there was no direct evidence showing its involvement in AD. Nowadays, scientists also borrow bioinformatics methods and tools to identity the role of lncRNAs in AD. Wang and co-workers used the propagation algorithm and identified two lncRNAs, H19 and PVT1, which might be associated with AD (Yang et al. 2014). ANRIL, the antisense non-coding RNA located in the INK4 locus (also called CDKN2B-AS), is also linked to Alzheimer's disease (Emanuele et al. 2011). The highly accelerated region 1 (HAR1) is located within a pair of overlapping lncRNAs, HAR1F, and HAR1R. The neurons of the human brain express both these lncRNAs (i.e., HAR1F and HAR1R). HAR1 is implicated in AD as well as other psychiatric diseases (Tamura et al. 2007). However, the role of these lncRNAs is still unclear in AD. Until now, scientists have predicted lncRNAs based on RNA sequencing data. The strategy to integrate RNA-sequencing technology and bioinformatics analysis will prove much more helpful to know about the mechanistic role of lncRNAs in such neural diseases.

LncRNA in Parkinson's disease

Parkinson's disease (PD), the second most serious neurological disorder, is characterized by slowness of movements, resting tremor, dementia, gait disturbance, and postural instability. Various lncRNAs found associated with this disorder are discussed below.

naPINK1 lncRNA

Parkinson's disease, characterized by the loss of brain dopaminergic neurons, is a chronic progressive movement disorder. The over-expression or loss of function of PTEN-induced kinase 1 (PINK1) has been implicated in the aberrant production of dopamine and impaired motor function (Chiba et al. 2009). NaPINK1 is a human-specific antisense lncRNA to PINK1. PINK1 is a mitochondrial kinase that protects cells against oxidative stress-induced apoptosis. NaPINK1 has been implicated in stabilizing PINK1 expression. Silencing naPINK1 results in the reduced expression of PINK1 expression in neurons (Scheele et al. 2007), thus promoting neuronal apoptosis and hence neurodegeneration. Therefore, a strategy to optimize levels of naPINK1 could be a good approach to tame PD.

AS-Uchl1 lncRNA

The role of the ubiquitin carboxyl-terminal hydrolase L1 gene (Uchl1) was earlier implicated in brain function. Previously, the loss of UCHL1 activity was studied in many neurodegenerative diseases (Carrieri et al. 2015). AS-Uchl1 is the antisense transcript of Uchl1 that can regulate the expression of Uchl1 at the post-transcriptional level and also promote contact between post-transcriptional mRNA and polysaccharides thereby enhancing the process of translation and promoting protein synthesis. Regulation of Uchl1 expression via AS-Uchl1 has been investigated as a new approach for the treatment of PD.

NEAT2 lncRNA

Nuclear-enriched abundant transcript-2 (NEAT2) is a highly conserved ncRNA that is greatly expressed in neuronal cells (Lipovich et al. 2012). Research shows that the over-expression of NEAT2 up-regulated α-synuclein expression, whereas the inhibition of NEAT2 down-regulated its expression. Administration of β-asarone, an important component of *Acorus tatarinowii* was found to decrease the levels of both NEAT2 and a-synuclein in the brain tissue in the PD mice model, thus indicating that β-asarone has potent anti-PD activity (Zhang et al. 2016). Therefore, searching for natural or synthetic compounds with the ability to reduce levels of NEAT2 lncRNA could prove productive in ameliorating PD.

HOTAIR lncRNA

The hox (homeobox) transcript antisense intergenic RNA (HOTAIR) lncRNA (approximately 2.2kb) is transcribed from the HOXC locus and is found highly expressed in Parkinson's disease. When up-regulated, HOTAIR can specifically increase the stability of LRRK2 mRNA. LRRK-2 is in turn implicated in the initiation/development of Parkinson's disease (Haelterman et al. 2014). Thus, HOTAIR may indirectly lead to the apoptosis of neurons by stabilizing LRRK2 mRNA and thus promoting Parkinson's disease (Wang et al. 2017). Therefore, targeting this lncRNA could prove an effective therapeutic strategy.

LncRNAs in autism spectrum disorder (ASD)

Differential expression of LncRNAs is also seen in a variety of other neurological disorders including autism spectrum disorder (ASD). On the basis of microarray assays, a total of 222 lncRNAs were found to exhibit aberrant expression in human ASD brains, compared with normal brains (Asuni et al. 2006). However, there is still a lack of information regarding the function of these putative ASD lncRNAs. Understanding the function of these lncRNAs could prove beneficial for understanding the molecular mechanism of ASD.

LncRNA in Huntington's disease

Huntington's disease (HD) is a progressive disorder of the brain characterized by uncontrolled movements, reduced thinking/reasoning abilities and emotional problems. The most common form of this disorder is adult-onset HD, which appears frequently in people in their 30s or 40s. This disorder is characterized by signs and symptoms including depression, irritability, poor coordination, small involuntary movements, poor learning, and declined ability to make decisions. HD patients develop involuntary jerking or twitching movements called chorea. These movements become more pronounced with the progress of the disease. Affected persons may have trouble speaking, walking, and swallowing. Patients suffering from the adult-onset form of HD usually live for fifteen to twenty years after the signs and symptoms become noticeable. Various lncRNAs implicated in HD are discussed below.

BDNFOS

The brain-derived neurotrophic factor (BDNF) is an important protein expressed extensively in the brain of adult mammals that promotes the survival of major neural cells affected in AD. BDNF is also implicated in the correct functioning of the corticostriatal synapse and the survival of the GABAergic medium-sized spiny striatal neurons that die in HD. BDNFOS is the antisense transcript of BDNF, which acts as a bodyguard and up-regulates the transcription of BDNF and the translation of BDNF-mRNA in HD. BDNFOS has a neuro-protective function and improves the HD phenotype (Wang et al. 2017). Thus, the strategy focusing on promoting the levels of BDNFOS could enhance the survival of brain cells and thus prove effective in ameliorating HD.

HttAS_v1 lncRNA

Huntington's disease (HD) characterized by progressive dyskinesias and mental decline involves the accumulation of Huntington (Htt) protein in neuronal cells that form large molecular aggregates in the brain leading to progressive death of the striatum and cortical nerve cells. HttAS_v1 LncRNA, the antisense transcript of the Htt gene, shows reduced expression in the frontal cortex of patients suffering from Huntington's disease, thereby leading to the enhanced expression of Htt mRNA and further aggravating the condition (Chung et al. 2011). Thus, a therapeutic strategy to promote the expression of HttAS_v1 could prove fruitful in ameliorating HD.

NEAT1 lncRNA

The nuclear paraspeckle assembly transcript 1(NEAT1) is an essential lncRNA implicated in the development and maintenance of paraspeckles (subnuclear bodies present in mammalian cells) (Clemson et al. 2009). A study by Sunwoo and colleagues showed an increased expression of NEAT1 in R6/2 mice and HD patients. In order to examine the effects of NEAT1 on the survival of neuronal cells, neuro2A cells were transfected with the NEAT1 short isoform vector and then subjected to hydrogen-peroxide induced damage. Neuro2A cells transfected with NEAT1 exhibited increased viability under oxidative stress, indicating that NEAT1 up-regulation has a role in protecting neurons against oxidative damage (Sunwoo et al. 2017).

HAR1F and HAR1 lncRNAs

One study examined the expression profile of ncRNAs in the brain tissue of HD patients and found that the expression of antisense transcripts of the first human accelerated region 1 (HAR1) gene, that is, lncRNAs-HAR1F and HAR1R, was significantly reduced in the striatum. These lncRNAs have roles in memory structure, neurotransmission, and synaptic plasticity in the mature brain and their abnormal expression pattern is also implicated in HD (Li et al. 2014).

MEG3 lncRNA

The aberrant nucleo-cytoplasmic exchange of REST in the striatum of the HD brain causes the transcriptional repression of HAR (Johnson et al. 2010). Another lncRNA, namely DGCR5, the transcript of DiGeorge critical region 5 (DGCR5), possesses a binding site for REST, which is an important regulator of transcription in HD (Johnson et al. 2009). The lncRNA MEG3 encoded by the maternally expressed gene 3 (MEG3) shows a down-regulated pattern in the HD brain that is inhibited by REST (Johnson 2012). Scientists believe that MEG3 might be involved in regulating the epigenetic profile of the developing and adult human brain, which in turn may alter the expression of various genes in response to neuronal activity. Therefore, the down-regulation of MEG3 lncRNA could probably possess serious consequences.

Abhd11os of lncRNA

Experiments in murine models of HD showed that the increased expression of the lncRNA Abhd11os (ABHD11-AS1 in humans) reduces the toxicity of Htt mRNA. The Abhd11os gene, when knocked out, produces neurotoxicity in the HD mouse model, whereas the over-expression of Abhd11os shows a neuroprotective effect (Francelle et al. 2015).

In general, it seems that lncRNAs operate in the pathology of HD via the following mechanisms: (1) by acting as a genetic enhancer/repressor (e.g., MEG3), (2) by acting as an epigenetic repressor (e.g., TUG1), (3) by showing antisense regulation (e.g., BDNF/BDNFOS), and (4) by acting as a transcriptional cofactor (e.g., Evf2).

LncRNAs in Angelman syndrome

Angelman syndrome is a complex genetic disorder of the nervous system, characterized by delayed development, severe speech impairment, intellectual disability, and problems with movement and balance (ataxia). Most children affected have a small head size (microcephaly) and experience recurrent seizures (epilepsy). Delayed development becomes evident by the age of 6 to 12 months; other common signs and symptoms generally appear in early childhood. Ubiquitin protein ligase E3A (UBE3A)-ATS is highly implicated in imprinting disorders like Angelman syndrome. This neuro-developmental disorder is caused by a defect in UBE3A. As a matter of fact, although UBE3A is expressed bi-allelically in most tissues, only the maternal allele contributes to UBE3A expression in the brain (Muddashetty et al. 2002). In the case of patients suffering from Angelman syndrome, the maternal UBE3A gene is inactivated or lost, thus resulting in the loss of functional UBE3A in neurons (Mus et al. 2007). The antisense lncRNA transcript UBE3A-ATS has been found to regulate epigenetic silencing of the paternal allele of UBE3A (Lewejohann et al. 2004) thus establishing the importance of this lncRNA in the pathology of this dreadful disease. On the basis of these lines, the strategy to target UBE3A-ATS could lead to the increased expression of paternal UBE3A in the brain and thus could constitute an effective treatment for Angelman syndrome.

LncRNA in amyotrophic lateral sclerosis (ALS)

ALS is characterized by the progressive degeneration of spontaneous motor neurons; it causes paralysis of limbs and the muscles involved in respiration, swallowing, and speech. In ALS, motor neurons die (atrophy) over time, leading to muscle weakness, a loss of muscle mass, and an inability to control movement. Presently, there is no effective treatment for this terrible neurodegenerative disease. Scientists reported the amplification of a six-nucleotide motif, “GGGGCC,” in the protein-coding gene C9ORF72 (chromosome 9 ORF 72) as the first causative mutation in ALS (Zu et al. 2013). Now, various non-coding transcripts have also been detected at the C9ORF72 locus. Since the transcription of the C9ORF72 repeats is bidirectional, it produces both sense and antisense RNA transcripts (Mizielinska et al. 2013). Both these transcripts are predominantly located in the nucleus of neurons, and show elevated levels in the brain of ALS patients (Lagier-Tourenne et al. 2013). C9ORF72 antisense lncRNA can inhibit C9ORF72 mRNA thus blocking gene expression. Fused in sarcoma or translated in liposarcoma (FUS/TLS) and TAR DNA-binding domain protein 43 (TDP43) are the proteins that bind RNA and that are mostly localized in the nucleus. These proteins play an important role in regulating the RNA metabolism. Aberrant accumulation of TDP43 and FUS/TLS is implicated in the misfolding of wild type Cu/Zn superoxide dismutase (wtSOD1) in non-SOD1 familial ALS (FALS) and sporadic ALS (SALS) patients, which indicates that they have a definite role in the pathogenesis of ALS (Pokrishevsky et al. 2012). Earlier reports show that an un-named lncRNA could recruit FUS/TLS to the genomic locus encoding for cyclin D1. Thus, transcription of cyclin D1 is repressed in response to DNA damage signals, leading to an increased tolerance of apoptotic signals. This evidence suggests that lncRNA may play a role in the pathogenesis of ALS (Wang et al. 2008).

Outlook and conclusion

LncRNAs represent potential targets for therapeutic intervention and drug development. Broadly speaking two approaches can be employed for developing lncRNA-based therapeutics, that is, lncRNA mimics and lncRNA antagonists. LncRNA are relatively short single strand oligonucleotides that can be easily synthesized *in vitro*. They exhibit excellent bioavailability and possess tissue half-lives of several weeks. Moreover, lncRNAs can be obtained through *in vitro* screening and then modified to improve their biological properties. All these properties make lncRNAs ideal candidates for developing drugs. The latest lncRNA information from GENCODE has overcome one of the biggest hurdles in lncRNA biology. Presently there is an urgency to progress in the functional and mechanistic aspects of the lncRNA field. The knockout animal models of lncRNAs could serve as the best tools to explore the roles of different lncRNAs in the human body, especially in the case of dreadful pathologies like neurological disorders. In reality, exploring the biological functions of lncRNAs is not an easy task and may pose complications because lncRNAs also function as precursors for small RNAs like small nucleolar RNAs (snoRNAs) or miRNAs (Derrien et al. 2012). Under such circumstances, displaying the recovery of the phenotype with the lncRNA in trans will be critical to determining whether the transcript functions as an lncRNA or acts merely as a precursor. Since many lncRNAs have been found to regulate epigenetic mechanisms, it will therefore be interesting to know where in the human genome they interact and what type of proteins they are shown interacting with. Techniques like RNA co-immunoprecipitation and deep sequencing (RIP-seq) can be helpful to explore the types of lncRNAs that show interaction with a particular protein. Alternatively, the pull-down method coupled with micro arrays can also be used to identify lncRNAs: through protein interactions. Examining the interactions of lncRNAs with RNA and proteins will provide clues and deep insights regarding the lncRNA function. Despite their pivotal role in human pathologies, only a small number of lncRNAs have so far been characterized. The latest ENCODE annotation of lncRNAs and other techniques for precipitation, purification, and knockdown of lncRNAs will provide a great service to the lncRNA field and smoothen the process of characterizing individual lncRNAs. Deep understanding of the mechanism of action of lncRNAs will help us explore the complex molecular interactions in which these lncRNAs participate and will also provide the basis for the development of effective drugs.

Conflict of interest

The authors declare that they have no conflicts of interest with respect to research, authorship, and publication of this book chapter.

Copyright and permission statement

We confirm that the materials included in this chapter do not violate copyright laws. Where relevant, appropriate permissions have been obtained from the original copyright holder(s). All original sources have been appropriately acknowledged and/or referenced.

Funding

No funding was received for this project.

References

Airaksinen, M. S., & Saarma, M. (2002). The GDNF family: signalling, biological functions and therapeutic value. *Nature Reviews Neuroscience*, *3*(5), 383–94.

Airavaara, M., Pletnikova, O., Doyle, M. E., Zhang, Y. E., Troncoso, J. C., & Liu, Q. R. (2011). Identification of novel GDNF isoforms and cis-antisense GDNFOS gene and their regulation in human middle temporal gyrus of Alzheimer disease. *Journal of biological chemistry*, *286*(52), 45093–102.

Andersen, O. M., Reiche, J., Schmidt, V., Gotthardt, M., Spoelgen, R., Behlke, J., . . . & Bales, K. R. (2005). Neuronal sorting protein-related receptor sorLA/LR11 regulates processing of the amyloid precursor protein. *Proceedings of the National Academy of Sciences*, *102*(38), 13461–66.

Assareh, A. A., Piguet, O., Lye, T. C., Mather, K. A., Broe, G. A., Schofield, P. R., . . . & Kwok, J. B. (2014). Association of SORL1 gene variants with hippocampal and cerebral atrophy and Alzheimer's disease. *Current Alzheimer research*, *11*(6), 558–63.

Asuni, A. A., Boutajangout, A., Scholtzova, H., Knudsen, E., Li, Y. S., Quartermain, D., . . . & Sigurdsson, E. M. (2006). Vaccination of Alzheimer's model mice with Aβ derivative in alum adjuvant reduces Aβ burden without microhemorrhages. *European journal of neuroscience*, *24*(9), 2530–42.

Berridge, M. J. (1998). Neuronal calcium signaling. *Neuron*, *21*(1), 13–26.

Bond, A. M., VanGompel, M. J., Sametsky, E. A., Clark, M. F., Savage, J. C., Disterhoft, J. F., & Kohtz, J. D. (2009). Balanced gene regulation by an embryonic brain ncRNA is critical for adult hippocampal GABA circuitry. *Nature neuroscience*, *12*(8), 1020.

Carrieri, C., Forrest, A. R., Santoro, C., Persichetti, F., Carninci, P., Zucchelli, S., & Gustincich, S. (2015). Expression analysis of the long non-coding RNA antisense to Uchl1 (AS Uchl1) during dopaminergic cells' differentiation *in vitro* and in neurochemical models of Parkinson's disease. *Frontiers in cellular neuroscience*, *9*, 114.

Cech, T. R., & Steitz, J. A. (2014). The noncoding RNA revolution—trashing old rules to forge new ones. *Cell*, *157*(1), 77–94.

Chiba, M., Kiyosawa, H., Hiraiwa, N., Ohkohchi, N., & Yasue, H. (2009). Existence of Pink1 antisense RNAs in mouse and their localization. *Cytogenetic and genome research*, *126*(3), 259–70.

Chung, D. W., Rudnicki, D. D., Yu, L., & Margolis, R. L. (2011). A natural antisense transcript at the Huntington's disease repeat locus regulates HTT expression. *Human molecular genetics*, *20*(17), 3467–77.

Clemson, C. M., Hutchinson, J. N., Sara, S. A., Ensminger, A. W., Fox, A. H., Chess, A., & Lawrence, J. B. (2009). An architectural role for a nuclear noncoding RNA: NEAT1 RNA is essential for the structure of paraspeckles. *Molecular cell*, *33*(6), 717–26.

Dinger, M. E., Amaral, P. P., Mercer, T. R., Pang, K. C., Bruce, S. J., Gardiner, B. B., . . . & Sunkin, S. M. (2008). Long noncoding RNAs in mouse embryonic stem cell pluripotency and differentiation. *Genome research*, *18*(9), 1433–45.

Duning, K., Buck, F., Barnekow, A., & Kremerskothen, J. (2008). SYNCRIP, a component of dendritically localized mRNPs, binds to the translation regulator BC200 RNA. *Journal of neurochemistry*, *105*(2), 351–59.

Emanuele, E., Lista, S., Ghidoni, R., Binetti, G., Cereda, C., Benussi, L., ... & Politi, P. (2011). Chromosome 9p21. 3 genotype is associated with vascular dementia and Alzheimer's disease. *Neurobiology of aging*, *32*(7), 1231–35.

Feng, J., Bi, C., Clark, B. S., Mady, R., Shah, P., & Kohtz, J. D. (2006). The Evf-2 noncoding RNA is transcribed from the Dlx-5/6 ultraconserved region and functions as a Dlx-2 transcriptional coactivator. *Genes & development*, *20*(11), 1470–84.

Francelle, L., Galvan, L., Gaillard, M. C., Petit, F., Bernay, B., Guillermier, M., . . . & Déglon, N. (2015). The striatal long noncoding RNA Abhd11os is neuroprotective against an N-terminal fragment of mutant huntingtin *in vivo*. *Neurobiology of aging*, *36*(3), 1601–e7.

García, N. V. D. M., Karayannis, T., & Fishell, G. (2011). Neuronal activity is required for the development of specific cortical interneuron subtypes. *Nature*, *472*(7343), 351–55.

Greer, P. L., & Greenberg, M. E. (2008). From synapse to nucleus: calcium-dependent gene transcription in the control of synapse development and function. *Neuron*, *59*(6), 846–60.

Guennewig, B., & Cooper, A. A. (2014). The central role of noncoding RNA in the brain. In *International review of neurobiology*, *116*, 153–94.

Haelterman, N. A., Yoon, W. H., Sandoval, H., Jaiswal, M., Shulman, J. M., & Bellen, H. J. (2014). A mitocentric view of Parkinson's disease. *Annual review of neuroscience*, *37*, 137–59.

Iyer, M. K., Niknafs, Y. S., Malik, R., Singhal, U., Sahu, A., Hosono, Y., . . . & Poliakov, A. (2015). The landscape of long noncoding RNAs in the human transcriptome. *Nature genetics*, *47*(3), 199.

Jin, C., Liu, X., Zhang, F., Wu, Y., Yuan, J., Zhu, J., . . . & Cheng, Z. (2013). An updated meta-analysis of the association between SORL1 variants and the risk for sporadic Alzheimer's disease. *Journal of Alzheimer's disease*, *37*(2), 429–37.

Johnson, R. (2012). Long non-coding RNAs in Huntington's disease neurodegeneration. *Neurobiology of disease*, *46*(2), 245–54.

Johnson, R., Richter, N., Jauch, R., Gaughwin, P. M., Zuccato, C., Cattaneo, E., & Stanton, L. W. (2010). Human accelerated region 1 noncoding RNA is repressed by REST in Huntington's disease. *Physiological genomics*, *41*(3), 269–74.

Johnson, R., Teh, C. H. L., Jia, H., Vanisri, R. R., Pandey, T., Lu, Z. H., . . . & Lipovich, L. (2009). Regulation of neural macroRNAs by the transcriptional repressor REST. *Rna*, *15*(1), 85–96.

Knauss, J. L., & Sun, T. (2013). Regulatory mechanisms of long noncoding RNAs in vertebrate central nervous system development and function. *Neuroscience*, *235*, 200–214.

Lagier-Tourenne, C., Baughn, M., Rigo, F., Sun, S., Liu, P., Li, H. R., . . . & Qiu, J. (2013). Targeted degradation of sense and antisense C9orf72 RNA foci as therapy for ALS and frontotemporal degeneration. *Proceedings of the National Academy of Sciences*, *110*(47), E4530–E4539.

Lessing, D. M. C., & Lee, J. T. (2013). X chromosome inactivation and epigenetic responses to cellular reprogramming. *Annual review of genomics and human genetics*, *14*, 85–110.

Lewejohann, L., Skryabin, B. V., Sachser, N., Prehn, C., Heiduschka, P., Thanos, S., . . . & Lipp, H. P. (2004). Role of a neuronal small non-messenger RNA: behavioural alterations in BC1 RNA-deleted mice. *Behavioural brain research*, *154*(1), 273–89.

Li, X., Wu, Z., Fu, X., & Han, W. (2014). lncRNAs: insights into their function and mechanics in underlying disorders. *Mutation research/reviews in mutation research*, *762*, 1–21.

Lin, L. F., Doherty, D. H., Lile, J. D., Bektesh, S., & Collins, F. (1993). GDNF: a glial cell line-derived neurotrophic factor for midbrain dopaminergic neurons. *Science*, *260*(5111), 1130–32.

Lipovich, L., Dachet, F., Cai, J., Bagla, S., Balan, K., Jia, H., & Loeb, J. A. (2012). Activity-dependent human brain coding/noncoding gene regulatory networks. *Genetics*, *192*(3), 1133–48.

Ma, Q. L., Galasko, D. R., Ringman, J. M., Vinters, H. V., Edland, S. D., Pomakian, J., . .. & Cole, G. M. (2009). Reduction of SorLA/LR11, a sorting protein limiting β-amyloid production, in alzheimer disease cerebrospinal fluid. *Archives of neurology*, *66*(4), 448–57.

Mercer, T. R., Qureshi, I. A., Gokhan, S., Dinger, M. E., Li, G., Mattick, J. S., & Mehler, M. F. (2010). Long noncoding RNAs in neuronal-glial fate specification and oligodendrocyte lineage maturation. *BMC neuroscience*, *11*(1), 14.

Mizielinska, S., Lashley, T., Norona, F. E., Clayton, E. L., Ridler, C. E., Fratta, P., & Isaacs, A. M. (2013). C9orf72 frontotemporal lobar degeneration is characterised by frequent neuronal sense and antisense RNA foci. *Acta neuropathologica*, *126*(6), 845–57.

Modarresi, F., Faghihi, M. A., Patel, N. S., Sahagan, B. G., Wahlestedt, C., & Lopez-Toledano, M. A. (2011). Knockdown of BACE1-AS nonprotein-coding transcript modulates beta-amyloid-related hippocampal neurogenesis. *International Journal of Alzheimer's Disease*, *2011*.

Muddashetty, R. S., Khanam, T., Kondrashov, A., Bundman, M., Iacoangeli, A., Kremerskothen, J., . . . & Brosius, J. (2002). Poly (A)-binding protein is associated with neuronal BC1 and BC200 ribonucleoprotein particles. *Journal of molecular biology*, *321*(3), 433–45.

Mus, E., Hof, P. R., & Tiedge, H. (2007). Dendritic BC200 RNA in aging and in Alzheimer's disease. *Proceedings of the National Academy of Sciences*, *104*(25), 10679–84.

Muslimov, I. A., Iacoangeli, A., Brosius, J., & Tiedge, H. (2006). Spatial codes in dendritic BC1 RNA. *The Journal of cell biology*, *175*(3), 427–39.

Ng, S. Y., Johnson, R., & Stanton, L. W. (2012). Human long non-coding RNAs promote pluripotency and neuronal differentiation by association with chromatin modifiers and transcription factors. *EMBO journal*, *31*(3), 522–33.

Onoguchi, M., Hirabayashi, Y., Koseki, H., & Gotoh, Y. (2012). A noncoding RNA regulates the neurogenin1 gene locus during mouse neocortical development. *Proceedings of the National Academy of Sciences*, *109*(42), 16939–44.

Parenti, R., Paratore, S., Torrisi, A., & Cavallaro, S. (2007). A natural antisense transcript against Rad18, specifically expressed in neurons and upregulated during β-amyloid-induced apoptosis. *European journal of neuroscience*, *26*(9), 2444–57.

Pokrishevsky, E., Grad, L. I., Yousefi, M., Wang, J., Mackenzie, I. R., & Cashman, N. R. (2012). Aberrant localization of FUS and TDP43 is associated with misfolding of SOD1 in amyotrophic lateral sclerosis. *PloS one*, *7*(4), e35050.

Ponjavic, J., Oliver, P. L., Lunter, G., & Ponting, C. P. (2009). Genomic and transcriptional co-localization of protein-coding and long non-coding RNA pairs in the developing brain. *PLoS genetics*, *5*(8), e1000617.

Ponjavic, J., Ponting, C. P., & Lunter, G. (2007). Functionality or transcriptional noise? Evidence for selection within long noncoding RNAs. *Genome research*, *17*(5), 556–65.

Ramos, A. D., Diaz, A., Nellore, A., Delgado, R. N., Park, K. Y., Gonzales-Roybal, G., Oldham, M. C., Song, J. S, & Lim, D. A. (2013). Integration of genome-wide approaches identifies lncRNAs of adult neural stem cells and their progeny *in vivo*. *Cell stem cell*, *12*(5), 616–28.

Scheele, C., Petrovic, N., Faghihi, M. A., Lassmann, T., Fredriksson, K., Rooyackers, O., Wahlestedt, C., Good, L., & Timmons, J. A. (2007). The human PINK1 locus is regulated *in vivo* by a non-coding natural antisense RNA during modulation of mitochondrial function. *BMC genomics*, *8*(1), 74.

Straten, G., Eschweiler, G. W., Maetzler, W., Laske, C., & Leyhe, T. (2009). Glial cell-line derived neurotrophic factor (GDNF) concentrations in cerebrospinal fluid and serum of patients with early Alzheimer's disease and normal controls. *Journal of Alzheimer's disease*, *18*(2), 331–37.

Struhl K. (2007). Transcriptional noise and the fidelity of initiation by RNA polymerase II. *Nature structural and molecular biology*, *14*(2), 103.

Sunwoo, J. S., Lee, S. T., Im, W., Lee, M., Byun, J. I., Jung, K. H., ... & Kim, M. (2017). Altered expression of the long noncoding RNA NEAT1 in Huntington's disease. *Molecular neurobiology*, *54*(2), 1577–86.

Tamura, Y., Kunugi, H., Ohashi, J., & Hohjoh, H. (2007). Epigenetic aberration of the human REELIN gene in psychiatric disorders. *Molecular psychiatry*, *12*(6), 593–600.

Tilgner, H., Knowles, D. G., Johnson, R., Davis, C. A., Chakrabortty, S., Djebali, S., . . . & Guigó, R. (2012). Deep sequencing of subcellular RNA fractions shows splicing to be predominantly co-transcriptional in the human genome but inefficient for lncRNAs. *Genome research*, *22*(9), 1616–25.

Uhde, C. W., Vives, J., Jaeger, I., & Li, M. (2010). Rmst is a novel marker for the mouse ventral mesencephalic floor plate and the anterior dorsal midline cells. *PLoS one*, *5*(1).

Ulitsky, I., Shkumatava, A., Jan, C. H., Sive, H., & Bartel, D. P. (2011). Conserved function of lincRNAs in vertebrate embryonic development despite rapid sequence evolution. *Cell*, *147*(7), 1537–50.

Wang, D. Q., Fu, P., Yao, C., Zhu, L. S., Hou, T. Y., Chen, J. G., . . . & Zhu, L. Q. (2018). Long non-coding RNAs, novel culprits, or bodyguards in neurodegenerative diseases. *Molecular Therapy: Nucleic Acids*, *10*, 269–76.

Wang, S., Zhang, X., Guo, Y., Rong, H., & Liu, T. (2017). The long noncoding RNA HOTAIR promotes Parkinson's disease by upregulating LRRK2 expression. *Oncotarget*, *8*(15), 24449.

Wang, X., Arai, S., Song, X., Reichart, D., Du, K., Pascual, G., ... & Kurokawa, R. (2008). Induced ncRNAs allosterically modify RNA-binding proteins in cis to inhibit transcription. *Nature*, *454*(7200), 126–30.

West, A. E., Chen, W. G., Dalva, M. B., Dolmetsch, R. E., Kornhauser, J. M., Shaywitz, A. J., . . . & Greenberg, M. E. (2001). Calcium regulation of neuronal gene expression. *Proceedings of the National Academy of Sciences*, *98*(20), 11024–31.

Wong, R. O., & Ghosh, A. (2002). Activity-dependent regulation of dendritic growth and patterning. *Nature reviews neuroscience*, *3*(10), 803–12.

Yang, X., Gao, L., Guo, X., Shi, X., Wu, H., Song, F., & Wang, B. (2014). A network based method for analysis of lncRNA-disease associations and prediction of lncRNAs implicated in diseases. *PloS one*, *9*(1).

Yoo, Y. M., Lee, C. J., & Kim, Y. J. (2012). Exogenous GDNF increase the migration of the neural stem cells with no protection against kainic acid-induced excitotoxic cell death in rats. *Brain research*, *1486*, 27–38.

Zhang, M., Cai, F., Zhang, S., Zhang, S., & Song, W. (2014). Overexpression of ubiquitin carboxyl-terminal hydrolase L1 (UCHL1) delays Alzheimer's progression *in vivo*. *Scientific reports*, *4*, 7298.

Zhang, Q. S., Wang, Z. H., Zhang, J. L., Duan, Y. L., Li, G. F., & Zheng, D. L. (2016). Beta-asarone protects against MPTP-induced Parkinson's disease via regulating long non-coding RNA MALAT1 and inhibiting α-synuclein protein expression. *Biomedicine & Pharmacotherapy*, *83*, 153–59.

Zhang, Z. (2016). Long non-coding RNAs in Alzheimer's disease. *Current topics in medicinal chemistry*, *16*(5), 511–19.

Zu, T., Liu, Y., Bañez-Coronel, M., Reid, T., Pletnikova, O., Lewis, J., . . . & Ostrow, L. W. (2013). RAN proteins and RNA foci from antisense transcripts in C9ORF72 ALS and frontotemporal dementia. *Proceedings of the National Academy of Sciences*, *110*(51), E4968–E4977.

CHAPTER 7

ROLE OF LONG NON-CODING RNAS IN HORMONAL DISORDERS WITH A FOCUS ON PCOS

MUDASAR NABI,[1]
SHAYAQ UL ABEER RASOOL,[2]
SAIRISH ASHRAF,[1] SHAJRUL AMIN[*1]

[1] Department of Biochemistry, University of Kashmir, Hazratbal, Srinagar-190006, India
[2] Department of Biotechnology, University of Kashmir, Hazratbal, Srinagar-190006, India
[*] Corresponding author
Shajrul Amin Associate Professor Department of Biochemistry University of Kashmir Hazratbal, Srinagar-190006
Tel.: 9419018174, E-mail: shajrulamin@uok.edu.in, shajrul@rediffmail.com

Abstract

Each gland of the endocrine system releases specific hormones into one's bloodstream. These hormones travel through one's blood to other cells and help control or coordinate many body processes. The endocrine system is a network of hormone-producing glands that help control most bodily functions like the metabolism, reproduction, growth, and development, and so on. A number of endocrine diseases arise in a human body as a result of imbalance in the production of endocrine hormones, most commonly diabetes mellitus, hypothyroidism, hyperthyroidism, prolactinoma, and polycystic ovary syndrome (PCOS). Among different endocrine disorders affecting females, PCOS is one of the most common endocrinopathy in pre-menopausal women. PCOS is a multifactorial, heterogeneous, complex genetic, endocrine, and metabolic disorder, diagnostically characterized by chronic anovulation, polycystic ovaries, and biochemical/clinical manifestations of hyperandrogenism. It is the most common endocrine disorder in premenopausal women with 5–10% prevalence worldwide. It is associated with increased risks of infertility, impaired glucose tolerance, type 2 diabetes mellitus, metabolic syndrome, cardiovascular disorders, and endometrial hyperplasia. The interactions of multiple inherited genetic factors related to hyperandrogenism and environmental or acquired factors, such as sedentary lifestyle can together trigger the dysregulation of androgen synthesis, which is the main factor causing ovarian follicles not to grow as much as a dominant follicle, resulting in oligo-/anovulation and PCOS. The pathogenesis of PCOS remains elusive, with contributions from insulin resistance, adipose tissue dysfunction, abnormal steroidogenesis, and hypothalamic–pituitary–ovarian dysregulation. Genetic variants and epigenetic environmental factors probably contribute to the dysregulation of these varied systems and raise new avenues of research investigation in the rapidly evolving field of PCOS. The coding sequences account for approximately 2% of the genome, while a large proportion of the genome is transcribed as non-coding RNAs, including microRNAs, small interfering RNAs, piwi-interacting RNAs, and long non-coding RNAs (lncRNAs). These non-coding RNAs have been proven to have important regulatory potential both in transcription and post-transcription events. LncRNAs are transcripts longer than 200 nucleotides without protein-coding capacity; they are emerging as crucial players in various biological processes via diverse mechanisms. LncRNAs can be intergenic (between protein-coding genes; lincRNA), natural antisense, or intronic transcripts, or can be transcribed from promoters and divergent enhancers. Accumulating evidence has shown that lncRNAs are key regulators of development, differentiation, cell proliferation, and apoptosis, and they may play critical roles in diverse human disorders, such as diabetes, cardiovascular disease, neurological disorders, and tumorigenesis. The recent discovery of lncRNAs has dramatically altered our understanding of human genetics. However, little is documented about the roles of lncRNAs in follicular development, steroidogenesis, and PCOS.

Keywords: PCOS, lncRNAs, pathogenesis, insulin resistance

Introduction

The endocrine system is a set of various glands that synthesize and secrete hormones, which control normal human physiological processes like growth and development, metabolism, sexual and reproductive functions, and so on. Endocrine hormones exert their affect by binding to the cell receptor present on distant target cells, resulting in a change in cellular functions. Production of these hormones and their secretion into the bloodstream must be tightly controlled to keep up the overall homeostasis of the human body. Any imbalance—either excessive or insufficient production of hormones—indicates a problem with the endocrine system, which results in an endocrine disorder. A number of endocrine diseases arise in the human body as a result of an imbalance in the production of endocrine hormones; the most common endocrine disorders include glucose homeostasis disorders such as diabetes mellitus, thyroid disorders (hypothyroidism and hyperthyroidism), obesity, polycystic ovary syndrome (PCOS), acromegaly (overproduction of growth hormone), Addison's disease (adrenal insufficiency), Cushing's syndrome (excessive cortisol production), pituitary disorders, Hashimoto's thyroiditis (an autoimmune disease resulting in the decreased production of thyroid hormone), and others. Endocrinopathies are classified into primary endocrinopathy, secondary endocrinopathy, and tertiary endocrinopathy. Primary endocrinopathy indicates problems with the action of the distal endocrine glands. Secondary endocrinopathy is indicative of abnormalities in the pituitary gland. Tertiary endocrine disorder is associated with the dysfunction of the hypothalamus and corresponding releasing hormones from hypothalamus. These endocrinopathies are often revealed by widespread symptoms, affecting multiple parts of the human body that can range in severity from mild to very severe depending upon the extent of the particular endocrinopathy. Treatment of various endocrinopathies depends on the particular disorder; however, it most often focuses on regulating the body's hormone balance using synthetic hormones.

Polycystic ovary syndrome (PCOS) is the most common endocrinopathy affecting approximately 4–8% of pre-menopausal women (Azziz et al. 2004). PCOS is primarily defined as a combination of anovulation, clinical or biochemical hyperandrogenism, and polycystic ovaries. The symptoms are heterogeneous but commonly include oligo/amenorrhea, hirsutism, acne, alopecia, acanthosis, anxiety, and obesity. PCOS has an early onset and typical clinical features develop during adolescence. PCOS is associated with an increased risk of oligo/anovulation, infertility, and obstetric complications like miscarriages, gestational diabetes, pre-eclampsia, and preterm labor (Vink et al. 2006). The etiology of the syndrome remains obscure and by the time affected females seek medical attention, the syndrome has become a self-perpetuating derangement where identification of the initiating factor has become difficult. Nonetheless, insulin resistance, adipose tissue dysfunction, abnormal steroidogenesis, hypothalamic–pituitary–ovarian dysregulation, genetic variants, and epigenetic environmental factors probably contribute to the dysregulation of various varied systems in PCOS.

The major portion of the human transcriptome is non-coding and these non-coding transcripts act as epigenetic modulators with gene regulatory potential both at transcriptional level and at post-transcriptional events. Non-coding RNAs are divided into short non-coding RNAs like microRNAs and long non-coding RNAs (lncRNAs). The best-studied short non-coding RNAs are microRNAs, varying between 15 to 24nts. These RNA molecules play a vital role in cellular functioning and their dysregulation has been linked to neurodegenerative diseases, cardiovascular diseases, and metabolic and reproductive disorders. LncRNAs, on the other hand, are non-coding transcripts with a length greater than 200 nucleotides; they are emerging as crucial players in various biological processes. Furthermore, there is evidence of the involvement of lncRNAs in a number of diseases, such as neurological disorders, cardiovascular disease, reproductive diseases, endocrine disorders, and in various cancers. LncRNAs have been found to play the lead role in a number of reproductive diseases, such as endometriosis and so on (Zhao et al. 2014). However, little is documented about the role of lncRNAs in PCOS and its associated conditions.

Definition and etiology of PCOS

PCOS was initially reported by American gynaecologists Irving F. Stein and Michael L. Leventhal in 1935 as the combination of hirsutism, amenorrhoea, chronic anovulation and infertility, obesity, and enlarged cystic ovaries (I. Stein and Leventhal 1935). The National Institutes of Health (NIH) in 1990 recommended the presence of hyperandrogenism and Oligo-ovulation as the main criteria for PCOS. The ESHRE/ASRM workshop (Rotterdam Criteria) in 2003 recommended that PCOS be defined when at least two of the following three features are present: (1) oligo- and/or anovulation, (2) hyperandrogenism (clinical and/or biochemical), and (3) polycystic ovaries. In 2006 the Androgen Excess and PCOS Society (AE–PCOS) considered PCOS to be a hyperandrogenism disorder and made it mandatory for diagnosis (Azziz et al. 2006).

PCOS has been studied at length over the years and various theories have been put forward to explain its etiology; however, its pathogenesis is still a mystery. Though each theory seemed relevant, none by itself explains the basic cause of PCOS.

Intrauterine theory predetermined that fetal overexposure to androgens alters the ovaries of the fetus, producing PCOS like a phenotype (Hague and Miller 1985). Evidence for early prenatal exposure to excess androgen leading to a PCOS-like phenotype in the progeny comes from various animal models.

The genetic theory earlier suggests that the genetics of PCOS is an autosomal dominant mode of its inheritance. Later it was confirmed that PCOS is an X-linked or polygenic disorder with a significant genetic influence on fasting insulin level, androstenediol glucuronide, and body mass index (BMI). A significant genetic and heritable component was demonstrated by familial and twin studies on PCOS. PCOS pathogenesis has been suggested to be multigenic; genes involved in steroid hormone synthesis and action, carbohydrate metabolism and fuel homeostasis, gonadotropin action, and the histocompatibility region have been implicated.

Clinical symptoms

The clinical features of PCOS are diverse and the presentation of symptoms varies across a person's life span. PCOS is associated with various reproductive, metabolic, and psychosocial complications. The symptoms start with reproductive and dermatological abnormalities, usually beginning in adolescence, then shifting over time to metabolic and psychological disorders.

(1) Ovulatory and menstrual dysfunction: PCOS presents as a broad spectrum of menstrual irregularities appearing often at puberty or later on during the reproductive years in women (Livadas et al. 2011). This is generally manifested as oligomenorrhea or secondary amenorrhea resulting from chronic oligoovulation/anovulation. Oligomenorrhea is defined as menses that occur at intervals of greater than 6 to 8 weeks (fewer than eight menstrual cycles per year or menses that occur at intervals greater than 35 days). Amenorrhea is defined as menses that occur at intervals of greater than 90 days (fewer than four menstrual cycles per year). Women diagnosed with PCOS have 70–80% menstrual dysfunction (Brassard 2008). Of women with oligo/amenorrhea, 90% have a chance of being diagnosed with PCOS, and up to 95% of PCOS-affected females have oligomenorrhea or amenorrhea (Kumarapeli 2008).

(2) Hyperandrogenism: Hyperandrogenism is the primary clinical manifestation of PCOS, characterized by abnormal levels of androgens (Labrie et al. 2011). Approximately 60 to 80% of PCOS-affected women show an increased concentration of circulating androgens like androstenedione and testosterone (Orio et al. 2003; Hahn et al. 2005). Clinical hyperandrogenism is manifested as hirsutism, acne, acanthosis, and alopecia (Lee and Zane 2006). Hirsutism, the most common clinical

manifestation of hyperandrogenism, is defined as excessive male pattern growth and distribution of terminal hair in women (Souter et al. 2004). High levels of circulating androgens are needed for hair growth on these body parts. Hirsutism is present in the majority of PCOS women and it also serves as a marker of excess androgen (Azziz et al. 2004). Acne is a chronic inflammation of the sebaceous glands that is also present on the faces of many adolescents (Archer and Chang 2004). Acne affects 15–25% of PCOS patients and besides the face can also be present on the neck, chest, and upper back. The main etiological factor is excess androgen because sebaceous gland activity is androgen dependent (Falsetti et al. 2002).

Androgenic alopecia is female pattern hair loss characterized by thinning (miniaturization) of the scalp hair growth with modest recession of the frontal hairlines (Olsen 2001). Alopecia has a reported wide range of prevalence, occurring in 3.2–34.8% of various populations of patients presenting for evaluation for PCOS; elevated metabolites of testosterone along with enhanced local 5α-RA activity are also reported in women with androgenic alopecia (Yildiz 2006).

The clinical manifestation of hyperandrogenism are mainly cosmetic, they give rise to emotional distress, anxiety, social embarrassment, and reduced quality of life and result in a psychosocial scream.

(3) Polycystic ovaries (PCO): The ESHRE/ASRM PCOS workshop in 2003 defined PCO as the presence of 12 or more follicles of 2–9 mm or an ovarian volume of $\geq$10 ml. The prevalence of polycystic ovaries is high among patients with PCOS ranging from 74.80–92.0% (Jonard et al. 2003).

Long-term effects of PCOS on health

PCOS has a serious impact on long-term health and increases the risk of reproductive, metabolic, and psychological complications (Daniilidis and Dinas 2009). Early complications like insulin resistance, obesity, dyslipidemia, and anovulation can culminate in long-term consequences like cardiovascular disorders, infertility, type 2 diabetes mellitus, and endometrial hyperplasia. The psychological complications include obstructive sleep apnea, non-alcoholic steatohepatitis, and psychiatric illnesses, such as depression, anxiety, and eating disorders, and a decrease in quality of life. The effect of immediate and long-term complications varies in severity because of the heterogeneous nature and the different PCOS phenotypes. The changes in phenotypes can change the clinical manifestations and the metabolic status can modify the expression of long-term complications (Welt and Carmina 2013).

Long non-coding RNAs

The genome sequencing by the human genome project provided quite a surprise with a fact that there are~ 20,000 protein-coding genes that correspond to approximately 2% of the total genomic sequence. What about the remaining 98% of the human genome? Earlier, it was presumed to be just non-functional junk DNA. The determination of the functional role of this ostensible junk DNA turned out to be the hotspot for researchers; the transcriptional complexity of the genome was revealed with the advancement in sequencing methodologies over the past decade. The traditional vision of the DNA transcription landscape and its mRNA-centric model for transcription annotation has undergone a fundamental change (Kapranov et al. 2002). The use of Cap analysis of gene expression (CAGE-sequencing) and genome-tiling arrays revealed that more of the genome is transcribed than was predicted earlier. The Encyclopedia of DNA Elements (ENCODE) project has approximated that transcription covers around 62–75% of the genome, most of which encodes mostly non-coding RNAs (ncRNAs) and that these non-coding RNAs have been found to play a significant role in almost every normal cellular and pathological processes (Djebali et al. 2012; ENCODE Project 2012). There is now increasing evidence that a major portion of the human genome undergoes transcription and that these transcripts lack the ability to act as a template for protein biosynthesis. These RNAs that lack the ability to act as templates for protein synthesis are referred to as ncRNAs. The part of the genome that does not code for proteins is transcribed to produce a varied range of non-coding RNAs. These non-coding RNAs are primarily divided into short non-coding RNAs (which have a transcript size of fewer than 200 nucleotides and encompass rRNA, miRNA, snRNA, snoRNAs, siRNA, and piRNA) and long non-coding RNAs with a transcript size exceeding 200 nucleotides. Following the discovery of ribosomal RNA and transfer RNA in the 1950s, biologically active non-coding RNAs started to be documented. Later, in the early 1990s, chromosome structure-regulating ncRNAs like XIST were discovered (Brockdorff et al. 1992). Since then, following extensive research, a large number of new biologically active ncRNAs have been discovered (Ulitsky and Bartel 2013; Rinn and Guttman 2014). This field of transcriptomics was the main focus for researchers, who found that nearly all the mammalian genome is transcribed at some stage with some individuals speculating that much of this pervasive transcription is likely functional (Djebali et al. 2012). The functional role of this cellular transcriptional output is yet to be fully elucidated and until now only a small number of ncRNAs like miRNAs, snoRNAs, and so on have been studied widely. Nevertheless, the non-coding component of the mammalian genome plays an important role in various diseases as well as in some critical biological processes such as development and differentiation (Ellis et al. 2014; Chen and Carmichael 2010). After being studied extensively since their discovery in 2001, miRNAs have provided essential insights about their role in the repression of target genes and in disease association. In fact, miRNA therapeutics have been tested in various clinical trials. After the explosion of interest in miRNAs, lncRNAs are now gaining most of the attention for their role in various biological processes. LncRNAs are non-coding transcripts with a length greater than 200 nucleotides. Like microRNAs, lncRNAs have surfaced as yet another key player in the regulation of gene expression.

Types and characteristics of LncRNAs

On the basis of their relative genomic position with respect to adjacent protein-coding genes, lncRNAs have been categorized into five types: these are intronic, enhancer lncRNAs, antisense, intergenic, and divergent (figure 1). Further, based on localization within the cell, lncRNAs are classified either as cytosolic or nuclear; nevertheless, some lncRNAs are present in both cellular compartments. LncRNAs present in a nucleus usually regulate expression of a target gene by altering chromatin structure, while cytosolic lncRNAs regulate expression of a target gene by affecting mRNA stability and translational efficiency through RNA–RNA interactions.

A small number of novel lncRNAs had been previously characterized, for example: the X-chromosome inactive-specific transcript (XIST) lncRNA, transcribed from the inactive X-chromosome in female mammals (Brockdorff et al. 1992) to regulate X-inactivation processes in cis; the imprinted transcript H19 (Sleutels et al. 2002), the antisense of insulin-like growth factor 2 non-coding RNA (Airn), which regulates the expression of insulin-like growth factor 2 (IGF2) in cis (Sleutels et al. 2002); and the lncRNA steroid receptor RNA activator (SRA), which enhances steroid hormone receptor responses (Lanz et al. 2002). However, most of the long non-coding RNAs remain uncharacterized and their functional role in the biological system remained unambiguous or was merely thought to be a byproduct of transcriptional noise. An additional 231 lncRNAs transcribed from homeobox (HOX) clusters were discovered in 2007. One of these lncRNAs, HOX transcript antisense RNA (HOTAIR), transcribed from the HOXC cluster, was shown to regulate coding transcripts from the HOXD cluster in trans, thus enlightening functions of lncRNAs beyond cis-regulation.

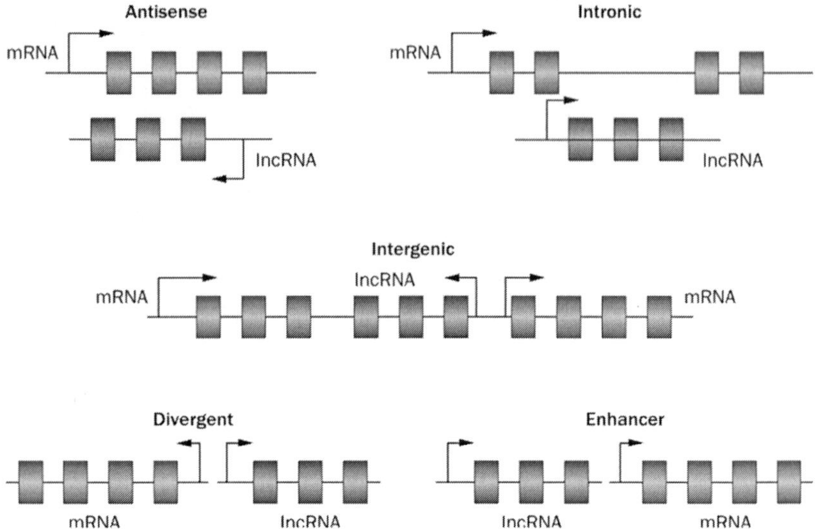

Figure 1. Adopted from Knoll et al. 2015, *Nat. Rev. Endocrinology*.

LncRNAs function as epigenetic modulators, they interact with chromatin-modifying proteins, targeting the catalytic activity of proteins towards a particular site on the chromatin to modify chromatin assembly and binding to target mRNAs to alter their expression levels. They can also act as competing endogenous RNA (ceRNA) of various microRNAs, thereby checking gene regulation. Furthermore, lncRNAs can also regulate gene expression by acting as competing endogenous RNAs (ceRNA) through microRNA competition (Mercer and Mattick 2013). With further development in the field of lncRNAs, new perspectives on their mechanism of action in gene regulation are expected in the near future.

LncRNAs involved in different hormonal disorders with a focus on PCOS

The critical roles of lncRNAs in normal cellular function and development has been established by a number of studies. The role of lncRNAs has been elucidated in various diseases like cancer, metabolic diseases, and other endocrine-related disorders such as diabetes mellitus and even in endocrine cancers. Studies have revealed that lncRNAs that have been involved in the development and regulation of the female reproductive system moreover play a crucial role in the pathogenesis of various gynecological-related diseases like endometriosis and various gynecological cancers such as cervical, endometrial, and ovarian (Zhao et al. 2014). Besides, they are also associated with some adverse pregnancy-related outcomes, like pre-eclampsia, spontaneous abortions, and preterm premature rupture of the fetal membranes (Wang et al. 2014). Furthermore, lncRNA have been shown to have autocrine, endocrine, and paracrine regulatory functions similar to that of miRNAs (Dragomir et al. 2018). In fact, some lncRNAs show autocrine-hormone-like behavior and can regulate cellular activity by controlling the transcription of a number of genes, such as genes encoding hormones or their receptors, and genes that code for hormone antagonists and thus directly or indirectly modify cellular activity and signaling transmission (Fatica et al. 2014). Studies have revealed that regulated expression of a number of lncRNAs in hormone response is required for these hormones to exert their full functions in various target cells. One of the first lncRNAs, the steroid receptor RNA activator (SRA), has been shown to be associated with hormone receptor pathways and to exhibit a hormone-like mechanism. SRA is expressed in tissues that express steroid receptors and hence are specifically targeted by steroid hormones, and acts as a co-activator of the steroid receptor to facilitate ligand-dependent transactivation (Lanz et al. 1999). Furthermore, *SRA* can also interact with co-repressors of nuclear receptors (Hatchell et al. 2006). A number of studies have revealed different lncRNAs that are involved in insulin resistance and the development of type 2 diabetes mellitus; the most important of these lncRNAs are natural antisense to PINK1 (naPINK1), ANRIL, H19, and HI-LNC25. The role of these lncRNAs has been well established in type 2 diabetes mellitus, one of most common endocrine disorders (Knoll et al. 2015). A number of lncRNAs have been shown to play a role in obesity, these include the functional intergenic repeating RNA element (FIRRE), SRA, adipogenic differentiation induced noncoding RNA (ADINR), and so on. The role of lncRNAs in the etiology of PCOS is still in its infancy. Some earlier studies carried out on Xist and Tsix lncRNAs that mediated the inactivation of the X-chromosome have provided the first important leads on the possible association of lncRNAs in PCOS pathophysiology. Nevertheless, research on the role of lncRNAs in the etiology of PCOS is expanding and a number of studies have been published already, which we will now discuss.

The H19 gene is a paternally imprinted gene located on chromosome 11p15.5, encoding a 2300nt-long lncRNA that plays an essential role in growth and development. This H19 lncRNA also regulates glucose homeostasis by competing for the endogenous RNA of miR-let-7 microRNA. It binds to miR-let-7 microRNA and prevents it from inhibiting its various target genes, including the insulin receptor gene and lipoprotein lipase gene (Kallen et al. 2013). H19 plays a significant role in the maintenance of the menstrual cycle and A. N. Kallen and co-workers observed down-regulation of various enzymes involved in steroidogenesis such as STAR and CYP11a1, likely mediated similarly to miR-let-7, when H19 was knocked down in the granulosa cells of PCOS patients. Subsequently, they also found a 2.8-fold increase in H19 expression in the cumulus cells of PCOS patients when compared with controls undergoing *in vitro* fertilization. Ovarian H19 possibly has an important role in the regulation of follicular development and androgen biosynthesis, which it is suggested play a key role in the development of PCOS

Another long non-coding RNA steroid receptor RNA activator (SRA) expressed by the Sra1gene is located on chromosome 5q31.3. LncRNA SRA regulates the activity of various transcription factors, including that of steroid receptors. In a study carried out by Zurvarra and co-workers (Zurvarra et al. 2009) in the letrozole-induced rat PCOS models, they observed up-regulation of androgen receptors and down-regulation of the estrogen and progesterone receptors in ovarian cells, thus indicating that steroid receptors play a vital role in the pathogenesis of PCOS. Further, the over-expression of the lncRNA SRA in PCOS women was found to be positively correlated with increased BMI and obesity (Liu et al. 2015a), suggesting it is a vital lncRNA for adiposity-related complications in PCOS (Liu et al. 2015a). Furthermore, the over-expression of SRA promotes cell growth, restrains cell apoptosis, and leads to the over-expression of estradiol and progesterone, and CYP19A1 and CYP11A1; the main candidates for PCOS thus signify that SRA may be a potential risk factor for PCOS (Li et al. 2018).

C-terminal binding protein 1 antisense (CTBP1-AS) lncRNA is a novel lncRNA that regulates the expression of the androgen receptor and also regulates androgen-receptor-related signaling pathways. The over-expression of CTBP1-AS lncRNA is presented as an independent risk factor for PCOS. In a study carried out by Z. Liu on peripheral blood leukocytes of PCOS women, CTBP1-AS was significantly over-expressed in these PCOS patients when compared with controls. Furthermore, the over-expression of CTBP1-AS was correlated with high testosterone and the insulin-resistance-marker homeostasis model of assessment for insulin resistance (HOMA-IR), values that thus suggest its role in PCOS-related hirsutism and insulin resistance (Liu et al. 2015b). Since CTBP1-AS lncRNA regulates AR activity and shows clinical significance in various diseases involving androgen receptors, this suggests its probable role in the pathophysiology of PCOS and PCOS-related hyperandrogenism.

LncRNAs in ovarian cumulus cells of PCOS are mostly up-regulated when compared with controls (Huang et al. 2016). In this study, 24,292 lncRNAs were examined in cumulus cells and a total of 623 lncRNAs were differentially regulated in PCOS patients with the fold change of ≥ 2. Of these 623, only 3 lncRNAs were down-regulated while 620 were up-regulated. Interestingly, the majority of the differentially expressed lncRNAs (n=74) were transcribed from chromosome 2. Chromosome number 2 is considered as an established genetic locus for PCOS since of 11 different susceptible loci for PCOS three are found on chromosome 2.

In 2017, Yu-dong Liu et al. carried out lncRNA expression profiles of mural granulosa cells. They first evaluated the differences between the lncRNA expression profiles of mural granulosa cells (GCs) from pre-ovulatory follicles after controlled ovarian hyperstimulation (COH) from PCOS patients and those from healthy women by microarray. A total of 862 lncRNAs were differentially regulated with a fold change of ≥ 2 in the PCOS granulosa cell compared with those from the non-PCOS group. Out of 862 differentially expressed lncRNAs, 692 were up-regulated and 170 were down-regulated. Then, to provide insight into the pathogenesis of PCOS by lncRNAs, they further investigated the possible role of differential lncRNA HLA complex group 26 (HCG26) in granulosa cell proliferation and steroidogenesis. It was found that lncRNA HCG26 is significantly more over-expressed in PCOS patients than in healthy controls and was associated with antral follicle count, suggesting the role of HCG26 lncRNA in cell proliferation in granulosa cells and contributing to excess antral follicles. They further found the knockdown of HCG26 in human granulosa cell line KGN decreased cell viability and induced G1/S arrest without having any effect on cellular apoptosis, suggesting that HCG26 knockdown restricts cell cycle progression thereby inhibiting cell proliferation. Further knockdown of HCG26 led to over-expression of the aromatase gene and enhanced estradiol (E2) production. The aromatase enzyme is encoded by the CYP19A1 gene. It converts androgens into estrogens thereby maintaining the homeostatic equilibrium between the two. CYP19A1 has been reported as one of the prime loci for the pathogenesis of PCOS. An increased testosterone level in the granulosa cell of PCOS patients is due to the decreased activity of the aromatase enzyme. This increased level of testosterone in the ovary promotes follicular growth and thus leads to PCOS morphology.

Lu-lu Fu and co-workers investigated the expression profiles of mRNAs and lncRNAs in the ovarian tissues of letrozole-induced PCOS rat model and control rats. In this study they found 158 lncRNAs and 2147 mRNAs were differentially expressed among the PCOS rat models and controls (Lu-lu Fu et al. 2018). Of 2147 differentially expressed mRNAs in the PCOS rat model, 1008 mRNAs were up-regulated and the remaining 1139 mRNAs were down-regulated. Furthermore, of a total 158 differentially expressed lncRNAs, 114 were up-regulated while 44 lncRNAs were down-regulated in the PCOS rat model. Next, they carried out gene ontology and expression analysis of these differentially expressed lncRNAs and mRNAs to determine the pathways they are involved in. From the gene ontology analysis, they found that aberrantly expressed lncRNAs and mRNAs were linked with reproduction, adhesion, and cellular metabolic processes. Furthermore, pathway analysis of these aberrantly expressed lncRNAs and mRNAs revealed that they were associated with several signal transduction pathways like steroidogenesis, insulin resistance, the proliferator-activated receptor (PPAR) signaling pathway, autoimmune thyroid disease, and the AMP-activated protein kinase (AMPK) signaling pathway. Among these signaling pathways, insulin resistance and the steroid hormone biosynthesis pathway have been reported to be associated with PCOS. These results indicate that these aberrantly expressed lncRNA/mRNA expressions may contribute to PCOS pathophysiology and may provide a clue to finding the therapeutic and diagnostic role of lncRNAs in PCOS.

The growth-arrest-specific transcript 5 (GAS5) lncRNA located on the 1q25 chromosomal region, 651 nt in size, regulates glucocorticoid signaling in metabolic and inflammatory pathways. The serum lncRNA GAS5 has been found to be down-regulated in PCOS women with IR when compared with PCOS subjects without IR and healthy controls (Haiyan et al. 2108). Furthermore, the area under the curve (AUC) model has demonstrated that GAS5 can be a good predictive biomarker in PCOS diagnosis. A number of earlier studies have shown that lncRNA GAS5 is involved in the cell cycle arrest in various diseases (Zhang et al. 2013). LncRNAs that regulate cell cycles might play a fundamental role in ovarian development and folliculogenesis, since PCOS is characterized by increased proliferation of the granulosa cell that contributes to abnormal folliculogenesis. GAS5 is thought to play an important role in the proliferation of the granulosa cell, folliculogenesis, and ultimately PCOS pathogenesis. Also, it is implied that GAS5 could involve the development of insulin resistance in PCOS.

Table 1. LncRNAs involved in PCOS.

S.No.	LncRNA	Gene location	Possible role in PCOS	Reference
1	H19	11p15.5	Down-regulates various enzymes involved in steroidogenesis like STAR and CYP11a1.	Kallen et al. 2013
2	SRA	5q31.3	Regulates activity of various androgen receptors and plays role in adiposity-related complications in PCOS.	Liu et al. 2015a
3	CTBP1-AS	4p16.3	Regulates the expression of androgen receptors and related signaling pathways. It also plays a role in PCOS-related hirsutism and insulin resistance.	Liu et al. 2015b
4	HCG26	6p21.33	HCG26 regulates expression of PCOS-related CYP genes and estradiol production.	Yu-dong Liu et al. 2017

Conclusion

LncRNAs have gained a lot of interest as potential new candidates involved in the regulation of basic cellular processes. LncRNAs have been implicated in a wide spectrum of developmental processes and molecular mechanisms governing normal physiological functioning. Thus, it is critical to understand their role in the cellular and molecular mechanisms of disease processes. Their discovery has led to a new understanding of the pathophysiology and progression of various diseases. They provide exciting prospects for using them in prognosis by targeting the specific pathways they regulate and hence providing a new prospectus in ameliorating various disorders. The lncRNAs can also be used as non-invasive diagnostic markers for various diseases.

PCOS is a multidimensional disorder involving metabolic, reproductive, and psychosocial complications. Its etiology remains a mystery but multifactorial interactions are proposed as a potential cause. LncRNAs are potential factors in the pathogenesis of an array of endocrine and metabolic disorders. Although lncRNA research in PCOS is still in its infancy, a limited number of studies that have been carried out show that they play a central role in understanding the genesis of PCOS and its associated conditions. LncRNAs can play a unifying role in understanding how genetic, epigenetic, and environmental factors can interact with each other, influence multiple regulatory mechanisms, bring about modifications in delicate metabolic endocrine balances in organisms, and lead to a wide spectrum of disorders.

Conflict of interest

The authors declare that they have no conflicts of interest with respect to research, authorship, and publication of this book chapter.

Copyright and permission statement

We confirm that the materials included in this chapter do not violate copyright laws. Where relevant, appropriate permissions have been obtained from the original copyright holder(s). All original sources have been appropriately acknowledged and/or referenced.

Funding

No funding was received for this project.

References

Archer, J. S., & Chang, R. J. (2004). Hirsutism and acne in polycystic ovary syndrome. *Best practice & research clinical obstetrics & gynaecology*, *18*(5), 737–54.

Azziz, R., Carmina, E., Dewailly, D., Diamanti-Kandarakis, E., Escobar-Morreale, H. F., Futterweit, W., . . . & Witchel, S. F. (2009). The Androgen Excess and PCOS Society criteria for the polycystic ovary syndrome: the complete task force report. *Fertility and sterility*, *91*(2), 456–88.

Azziz, R., Sanchez, L. A., Knochenhauer, E. S., Moran, C., Lazenby, J., Stephens, K. C., . . . & Boots, L. R. (2004). Androgen excess in women: experience with over 1000 consecutive patients. *Journal of clinical endocrinology & metabolism*, *89*(2), 453–62.

Brassard, M., AinMelk, Y., & Baillargeon, J. P. (2008). Basic infertility including polycystic ovary syndrome. *Medical clinics of North America*, *92*(5), 1163–92.

Brockdorff, N., Ashworth, A., Kay, G. F., McCabe, V. M., Norris, D. P., Cooper, P. J., . . . & Rastan, S. (1992). The product of the mouse Xist gene is a 15 kb inactive X-specific transcript containing no conserved ORF and located in the nucleus. *Cell*, *71*(3), 515–26.

Daniilidis, A., & Dinas, K. (2009). Long term health consequences of polycystic ovarian syndrome: a review analysis. *Hippokratia*, *13*(2), 90.

Djebali, S., Davis, C. A., Merkel, A., Dobin, A., Lassmann, T., Mortazavi, A., . . . & Xue, C. (2012). Landscape of transcription in human cells. *Nature*, *489*(7414), 101–8.

Doi, S. A. (2008). Neuroendocrine dysfunction in PCOS: a critique of recent reviews. *Clinical medicine & research*, *6*(2), 47–53.

Dragomir, M., Chen, B., & Calin, G. A. (2018). Exosomal lncRNAs as new players in cell-to-cell communication. *Translational cancer research*, *7*(Suppl 2), S243.

Ellis, B. C., Graham, L. D., & Molloy, P. L. (2014). CRNDE, a long non-coding RNA responsive to insulin/IGF signaling, regulates genes involved in central metabolism. *Biochimica et biophysica acta (BBA)-molecular cell research*, *1843*(2), 372–86.

ENCODE Project Consortium. (2012). An integrated encyclopedia of DNA elements in the human genome. *Nature*, *489*(7414), 57–74.

ESHRE, T. R., & ASRM-Sponsored PCOS Consensus Workshop Group. (2004). Revised 2003 consensus on diagnostic criteria and long-term health risks related to polycystic ovary syndrome. *Fertility and sterility*, *81*(1), 19–25.

Falsetti, L., Gambera, A., Andrico, S., & Sartori, E. (2002). Acne and hirsutism in polycystic ovary syndrome: clinical, endocrine–metabolic and ultrasonographic differences. *Gynecological endocrinology*, *16*(4), 275–84.

Fatica, A., & Bozzoni, I. (2014). Long non-coding RNAs: new players in cell differentiation and development. *Nature Reviews Genetics*, *15*(1), 7–21.

Fu, L. L., Xu, Y., Li, D. D., Dai, X. W., Xu, X., Zhang, J. S., . . . & Zheng, L. W. (2018). Expression profiles of mRNA and long noncoding RNA in the ovaries of letrozole-induced polycystic ovary syndrome rat model through deep sequencing. *Gene*, *657*, 19–29.

Hague, W. M., & Millar, D. R. (1985). Excessive testosterone secretion in pregnancy. Case report. *BJOG: International journal of obstetrics & gynaecology*, *92*(2), 173–78.

Hahn, S., Tan, S., Elsenbruch, S., Quadbeck, B., Herrmann, B. L., Mann, K., & Janssen, O. E. (2005). Clinical and biochemical characterization of women with polycystic ovary syndrome in North Rhine-Westphalia. *Hormone and metabolic research*, *37*(07), 438–44.

Hatch, R., Rosenfield, R. L., Kim, M. H., & Tredway, D. (1981). Hirsutism: implications, etiology, and management. *American journal of obstetrics and gynecology*, *140*(7), 815–30.

Hatchell, E. C., Colley, S. M., Beveridge, D. J., Epis, M. R., Stuart, L. M., Giles, K. M., . . . & Arthur, P. G. (2006). SLIRP, a small SRA binding protein, is a nuclear receptor corepressor. *Molecular cell*, *22*(5), 657–68.

Huang, X., Hao, C., Bao, H., Wang, M., & Dai, H. (2016). Aberrant expression of long noncoding RNAs in cumulus cells isolated from PCOS patients. *Journal of assisted reproduction and genetics*, *33*(1), 111–21.

Jonard, S., Robert, Y., Cortet-Rudelli, C., Pigny, P., Decanter, C., & Dewailly, D. (2003). Ultrasound examination of polycystic ovaries: is it worth counting the follicles? *Human reproduction*, *18*(3), 598–603.

Kallen, A. N., Karakaya, C., Seli, E., & Huang, Y. (2013). H19 expression is increased in cumulus cells of PCOS patients undergoing IVF. *Fertility and sterility*, *100*(3), S346.

Kapranov, P., Cawley, S. E., Drenkow, J., Bekiranov, S., Strausberg, R. L., Fodor, S. P., & Gingeras, T. R. (2002). Large-scale transcriptional activity in chromosomes 21 and 22. *Science*, *296*(5569), 916–19.

Kashar-Miller, M., & Azziz, R. (1999). Heritability and the risk of developing androgen excess. *Journal of steroid biochemistry and molecular biology*, *69*(1–6), 261–68.

Knoll, M., Lodish, H. F., & Sun, L. (2015). Long non-coding RNAs as regulators of the endocrine system. *Nature Reviews Endocrinology*, *11*(3), 151.

Kumarapeli, V., Seneviratne, R. D. A., Wijeyaratne, C. N., Yapa, R. M. S. C., & Dodampahala, S. H. (2008). A simple screening approach for assessing community prevalence and phenotype of polycystic ovary syndrome in a semiurban population in Sri Lanka. *American journal of epidemiology*, *168*(3), 321–28.

Labrie, F., Martel, C., & Balser, J. (2011). Wide distribution of the serum dehydroepiandrosterone and sex steroid levels in postmenopausal women: role of the ovary? *Menopause*, *18*(1), 30–43.

Lanz, R. B., McKenna, N. J., Onate, S. A., Albrecht, U., Wong, J., Tsai, S. Y., ... & O'Malley, B. W. (1999). A steroid receptor coactivator, SRA, functions as an RNA and is present in an SRC-1 complex. *Cell*, *97*(1), 17–27.

Lanz, R. B., Razani, B., Goldberg, A. D., & O'Malley, B. W. (2002). Distinct RNA motifs are important for coactivation of steroid hormone receptors by steroid receptor RNA activator (SRA). *Proceedings of the National Academy of Sciences*, *99*(25), 16081–86.

Lee, A. T., & Zane, L. T. (2007). Dermatologic manifestations of polycystic ovary syndrome. *American journal of clinical dermatology*, *8*(4), 201–19.

Legro, R. S., Driscoll, D., Strauss, J. F., Fox, J., & Dunaif, A. (1998). Evidence for a genetic basis for hyperandrogenemia in polycystic ovary syndrome. *Proceedings of the National Academy of Sciences*, *95*(25), 14956–60.

Li, Y., Wang, H., Zhou, D., Shuang, T., Zhao, H., & Chen, B. (2018). Up-regulation of long noncoding RNA SRA promotes cell growth, inhibits cell apoptosis, and induces secretion of estradiol and progesterone in ovarian granular cells of mice. *Medical science monitor: international medical journal of experimental and clinical research*, *24*, 2384.

Lin, H., Xing, W., Li, Y., Xie, Y., Tang, X., & Zhang, Q. (2018). Downregulation of serum long noncoding RNA GAS5 may contribute to insulin resistance in PCOS patients. *Gynecological endocrinology*, *34*(9), 784–88.

Lindgren, C. M., & McCarthy, M. I. (2008). Mechanisms of disease: genetic insights into the etiology of type 2 diabetes and obesity. *Nature clinical practice endocrinology & metabolism*, *4*(3), 156–63.

Liu, Y. D., Li, Y., Feng, S. X., Ye, D. S., Chen, X., Zhou, X. Y., & Chen, S. L. (2017). Long noncoding RNAs: potential regulators involved in the pathogenesis of polycystic ovary syndrome. *Endocrinology*, *158*(11), 3890–99.

Liu, Z., Hao, C., Huang, X., Zhang, N., Bao, H., & Qu, Q. (2015). Peripheral blood leukocyte expression level of lncRNA steroid receptor RNA activator (SRA) and its association with polycystic ovary syndrome: a case control study. *Gynecological Endocrinology*, *31*(5), 363–68.

Liu, Z., Hao, C., Song, D., Zhang, N., Bao, H., & Qu, Q. (2015). Androgen receptor coregulator CTBP1-AS is associated with polycystic ovary syndrome in Chinese women: a preliminary study. *Reproductive sciences*, *22*(7), 829–37.

Livadas, S., Christou, M., Economou, F., Karachalios, A., Xyrafis, X., Boutzios, G., . . . & Diamanti-Kandarakis, E. (2011). Menstrual irregularities in PCOS: does it matter when it starts? *Experimental and clinical endocrinology & diabetes*, *119*(06), 334–37.

Mercer, T. R., & Mattick, J. S. (2013). Structure and function of long noncoding RNAs in epigenetic regulation. *Nature structural & molecular biology*, *20*(3), 300.

Merino, P. M., Codner, E., & Cassorla, F. (2011). A rational approach to the diagnosis of polycystic ovarian syndrome during adolescence. *Arquivos Brasileiros de endocrinologia & metabologia*, *55*(8), 590–98.

Norman, R. J., Dewailly, D., Legro, R. S., & Hickey, T. E. (2007). Polycystic ovary syndrome. *Lancet*, *370*(9588), 685–97.

Olsen, E. A. (2001). Female pattern hair loss. *Journal of the American Academy of Dermatology*, *45*(3), S70–S80.

Orio Jr, F., Matarese, G., Di Biase, S., Palomba, S., Labella, D., Sanna, V., . . . & Lombardi, G. (2003). Exon 6 and 2 peroxisome proliferator-activated receptor-γ polymorphisms in polycystic ovary syndrome. *Journal of clinical endocrinology & metabolism*, *88*(12), 5887–92.

Palomba, S., Santagni, S., Falbo, A., & La Sala, G. B. (2015). Complications and challenges associated with polycystic ovary syndrome: current perspectives. *International journal of women's health*, *7*, 745.

Pekonen, F., Laatikainen, T., Buyalos, R., & Rutanen, E. M. (1989). Decreased 34K insulin-like growth factor binding protein in polycystic ovarian disease. *Fertility and sterility*, *51*(6), 972–75.

Rinn, J., & Guttman, M. (2014). RNA and dynamic nuclear organization: Long noncoding RNAs may function as organizing factors that shape the cell nucleus. *Science (New York, NY)*, *345*(6202), 1240.

Rotterdam ESHRE/ASRM-Sponsored PCOS Consensus Workshop Group. (2004). Revised 2003 consensus on diagnostic criteria and long-term health risks related to polycystic ovary syndrome (PCOS). *Human reproduction*, *19*(1), 41–47.

Sleutels, F., Zwart, R., & Barlow, D. P. (2002). The non-coding Air RNA is required for silencing autosomal imprinted genes. *Nature*, *415*(6873), 810–13.

Ulitsky, I., & Bartel, D. P. (2013). lincRNAs: genomics, evolution, and mechanisms. *Cell*, *154*(1), 26–46.

Vink, J. M., Sadrzadeh, S., Lambalk, C. B., & Boomsma, D. I. (2006). Heritability of polycystic ovary syndrome in a Dutch twin-family study. *Journal of clinical endocrinology & metabolism*, *91*(6), 2100–104.

Wang, H., Cao, Q., Ge, J., Liu, C., Ma, Y., Meng, Y., . . . & Wang, Y. (2014). LncRNA-regulated infection and inflammation pathways associated with pregnancy loss: genome-wide differential expression of lncRNAs in early spontaneous abortion. *American journal of reproductive immunology*, *72*(4), 359–75.

Welt, C. K., & Carmina, E. (2013). Lifecycle of polycystic ovary syndrome (PCOS): from in utero to menopause. *Journal of clinical endocrinology & metabolism*, *98*(12), 4629–38.

Yildiz, B. O. (2006). Diagnosis of hyperandrogenism: clinical criteria. *Best practice & research clinical endocrinology & metabolism*, *20*(2), 167–76.

Zhang, Z., Zhu, Z., Watabe, K., Zhang, X., Bai, C., Xu, M., . . . & Mo, Y. Y. (2013). Negative regulation of lncRNA GAS5 by miR-21. *Cell death & differentiation*, *20*(11), 1558–68.

Zhao, Y., Yang, Y., Trovik, J., Sun, K., Zhou, L., Jiang, P., . . . & Wang, H. (2014). A novel wnt regulatory axis in endometrioid endometrial cancer. *Cancer research*, *74*(18), 5103–17.

Zurvarra, F. M., Salvetti, N. R., Mason, J. I., Velazquez, M. M., Alfaro, N. S., & Ortega, H. H. (2009). Disruption in the expression and immunolocalisation of steroid receptors and steroidogenic enzymes in letrozole-induced polycystic ovaries in rats. *Reproduction, fertility and development*, *21*(7), 827–39.

CHAPTER 8

DIVERSE FUNCTIONS OF LONG NON-CODING RNAS IN CARDIOVASCULAR DISEASE

BILAL AHMAD TANTRY,[*1] QAMAR TABAN,[2,6]
MUDASAR NABI,[4]
PEERZADA TAJAMUL MUMTAZ[5,6]
SHAIK RAHIMAN,[3]
AND MOHAMMAD ABDUL HAFEEZ[1]

[1] Department of Microbiology, College of Medicine, Jouf University, Al-Jouf, 74331, Saudi Arabia
[2] Drug Therapeutics and Neurobiology Lab, Department of Biotechnology, School of Biological Sciences, University of Kashmir, Hazratbal, Srinagar, Jammu and Kashmir-190006, India
[3] Department of Biochemistry, College of Medicine, Jouf University, Al-Jouf, 74331, Saudi Arabia
[4] Department of Biochemistry, School of Biological Sciences, University of Kashmir, Srinagar, J&K-190006, India
[5] Department of Biochemistry, School of Life Sciences, Jaipur National University, Jaipur, Rajasthan-302017, India
[6] Division of Biotechnology, Faculty of Veterinary Sciences and Animal Husbandry, Alusteng, Shuhama, SKUAST-Kashmir-190006, India
*Corresponding author:
Dr. Bilal Tantry
Assistant Professor, Department of Microbiology, College of Medicine, Jouf University
Tel: +966506281724; E-mail: batantry@ju.edu.sa, bilaltantry@gmail.com

Abstract

Despite significant improvements over the past three decades, cardiovascular disease remains a leading worldwide health epidemic. The recent identification of a fascinating group of mediators known as long non-coding RNAs (lncRNAs) has provided a wealth of new biology to explore cardiovascular risk mitigation. LncRNAs are expressed in a highly context-specific fashion, and multiple lines of evidence implicate them in diverse biological processes. Indeed, abnormalities of lncRNAs have been directly linked with human ailments, including cardiovascular biology and disease. Of particular interest to the cardiovascular research community, dysregulation in lncRNA regulatory circuits has been associated with cardiac pathological hypertrophy, vascular disease, cell-fate programming and development, atherosclerosis, dyslipidemia, and metabolic syndrome. In this chapter, we discuss the multifaceted functions of long non-coding RNAs and discuss the current literature with an emphasis on cardiac development and disease. Furthermore, as the enormous spectrum of long non-coding RNAs potentially opens up new avenues for diagnosis and prevention of heart failure, we ultimately also evaluate the future prospects for long non-coding RNAs as biomarkers, and therapeutic targets for the treatment of cardiovascular disorders.

Keywords: lncRNAs, cardiovascular diseases, CAD, myocardial infarction, heart

Introduction

Heart and circulatory disorders are considered to be the types of disease that are the major leading cause for morbidity and mortality around the world. Globally, cardiovascular diseases (CVD) have become a leading health care burden issue. As per a WHO report (WHO 2012), about 46.2% (17.5 million) deaths around the world were the result of CVD in 2012 and this is expected to reach about 22.3 million in 2030. As per a BHF cardiovascular statistics (2014) report, the most prominent CVD is coronary heart disease, which accounts for 46% of global CVD deaths. The severe effects of coronary heart disease are myocardial infarction (MI) and damage to cardiac muscles due to hypoxia; atherosclerotic plaque in the vessel walls are the main causes of such diseases, and the condition can be worsened if the plaque ruptures and results in thrombosis (Danesh et al. 2004). Vasculogenesis (nascent) and angiogenesis (angioblasts) are the two main processes that lead to the formation of blood vessels. Vascular homeostasis is regulated by vascular endothelium and possesses metabolic, synthetic, immunological, and secretory functions (Aird 2004).

In the human body, the endothelial cells in a monolayer pattern line every blood vessel and are associated with cell-to-cell, gap, tight, and adherence junctions (Bazzoni and Dejana 2004). The passage of small molecular weight solutes takes place between two adjacent cells through the gap junction. The barrier between circulating blood vessels is through tight junctions and adherence junctions that regulate endothelial permeability to solutes and the leucocytes (Wallez and Huber 2008). Nitric oxide and prostaglandins act as mediators (anti-thrombotic) that are secreted by the endothelial surface, which controls thrombosis, coagulation, and vascular tone (West et al. 2001).

Non-coding RNA (ncRNA)

High-throughput transcriptomic analysis has revealed that only 1–2% of genomic DNA from eukaryotic genomes codes for proteins and the remaining 90% are transcribed as non-coding RNAs (ncRNAs) (Consortium 2004). The expression of ncRNAs occurs in a developmental stage-specific manner or cell/tissue-specific manner and dominates the genomes of protein-coding genes in multicellular organisms (van Rooij 2011; Ward et al. 2015). There is mounting evidence that ncRNAs are involved in several human diseases, such as cardiovascular, malignant, and neurological diseases, and certain growth-related disease mechanisms (Esteller 2011). Regulatory non-coding RNAs are categorized into small interfering RNAs (siRNAs), piwi-interacting RNAs (piRNAs), microRNAs (miRNAs), long non-coding RNAs (lncRNAs), enhancer RNAs (eRNAs), promoter-associated RNAs (PARs), and circular RNAs (circRNAs) (Cech and Steitz 2014).

Long non-coding RNAs (lncRNAs)

In recent years, long non-coding RNAs (lncRNAs) have gained attention as key regulators of cellular function and diseases. LncRNAs are greater than 200 nt in length and show classical mRNA similarities throughout the genome with high functional versatility in the regulation of gene expressions. LncRNAs are currently defined as RNA transcripts with no potential to encode for functional proteins, which separate them from miRNAs and from protein-coding genes (Rinn and Chang 2012). In the human genome, more than one hundred thousand lncRNA genes have been identified, which suggests a dominant role for lncRNA genes in the mammalian genome (O'Donnell et al. 2005).

The biological functions of most lncRNAs are currently still unknown, as only a few lncRNAs have been studied in detail. LncRNAs can be classified into different subtypes on the basis of their genomic location: sense lncRNAs, antisense lncRNAs, bidirectional lncRNAs, intronic lncRNAs, and intergenic lncRNAs (figure 1) (Thum and Condorelli 2015). Accumulating evidence suggests that lncRNAs play a significant role in a wide variety of important biological processes including regulating gene transcription, splicing, translation, cell cycle, apoptosis, cell differentiation, stem cell pluripotency, and heat shock response (Loewer et al. 2010). LncRNAs play important roles in many diseases like cancer (Sahu et al. 2015; Huarte 2015) and heart-related diseases (Uchida and Dimmeler 2015; Lorenzen and Thum 2016). However, only a few studies have revealed the role of lncRNAs in PAH (pulmonary arterial hypertension) and the expression of lncRNAs in rat lung tissue following hypoxia-mediated induction of PAH (Wang et al. 2016).

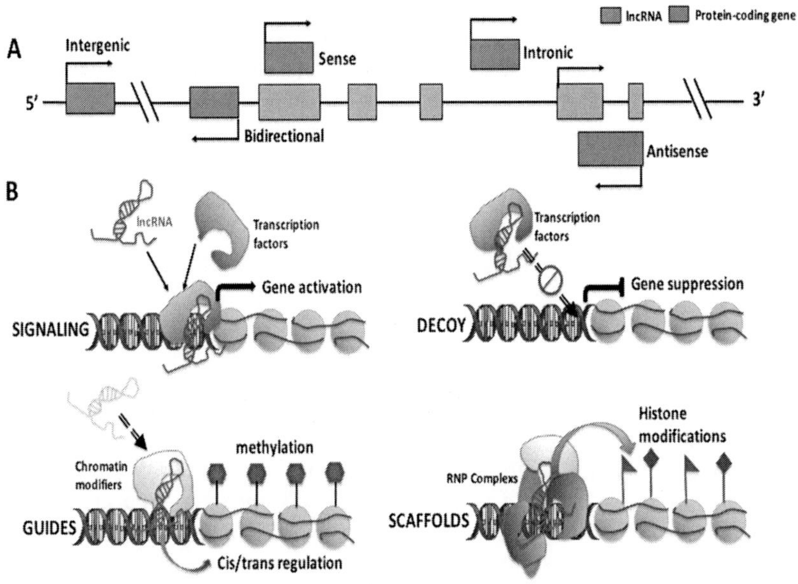

Figure 1. LncRNAs genome location and functions. Reproduced from Deng et al., 2015, Circ Res.

Transcription of lncRNA is shown in figure 1(A). The lncRNA transcripts are classified as sense, antisense, intronic, and intergenic or bidirectional. The main functions of lncRNA are shown in figure 1(B): Signaling: lncRNA act as a signal molecule or promote transcriptional activity. Decoy: lncRNAs are involved in access to regulatory proteins to DNA. Guides: lncRNAs can be used in protein complex localization either in cis or in trans. Scaffolds: lncRNAs can act as adaptors to form discrete complexes.

Through microarray analysis, 362 lncRNAs have been revealed by RT-PCR in hypoxic animal (normoxic) controls (Wang et al. 2016). A related study performed microarray analysis on endothelial tissues from the pulmonary arteries of chronic thromboembolic pulmonary hypertension (CETPH) patients and healthy controls, identifying 185 differentially-expressed lncRNAs observed in the CTEPH tissues compared with healthy controls. Gene ontology and KEGG pathway analysis revealed the role of lncRNAs in the regulation of inflammation, antigen presentation, and processing and responses to endogenous stimuli (Gu et al. 2015). In addition, RNA-seq analyzed the expression of lncRNAs in human heart failure, which is the primary cause of death in patients with PAH. Compared with heart donors, there are 105 lncRNAs expressed differentially in heart failures (Di Salvo et al. 2015). The expression profiles of lncRNAs in animal models and human tissues show that lncRNAs may participate in the pathogenesis of PAH.

Applications of lncRNAs in cardiovascular biology

Currently, several lncRNAs have been identified and their correlations with cardiovascular diseases (CVD) were reported. The following are some of the lncRNAs associated with CVD pathogeneses in humans as well as rodents.

In humans, differentiation models (ESCs) provide a powerful tool for the discovery and functional identification of novel lncRNAs. LncRNAs like Terminator, Alien, and Punisher play an important role at different stages of mesodermal and cardiovascular development. Terminator is an abundant lncRNA transcript in stem cells that plays an essential role in early embryonic survival and pluripotency both in zebrafish and mouse embryos. In contrast, Alien is more specifically expressed in mesendodermal tissues and leads to disrupted vasculature and heart formation if its function is lost. Interestingly, Alien is also an endodermal lncRNA participating in pancreatic differentiation through its interaction with Foxa2 and Smad2/3 (Jiang et al. 2015).

LncRNA (Bvht)

LncRNA Braveheart (*Bvht*) was the first lncRNA identified to be involved as a regulator of cardiac development. It is expressed in mouse embryonic stem cells (ESCs) and is known to act as a transitional regulator for the process of nascent mesoderm to cardiac progenitor states. Braveheart is a mouse-specific lncRNA; no homologs have yet been found in other species.

Braveheart AGIL motif is necessary for proper ESC differentiation using 11 nt deletion in AGIL via CRISPR/Cas9 (Xue et al., 2016).

LncRNA (Fendrr)

The tissue-specific lncRNA fetal-lethal non-coding developmental regulatory RNA (*Fendrr*) is shown to be involved in attenuating the differentiation of mesoderm and has a role in the formation of the heart and body wall in mice via modulating PRC2 or TrxG/MLL activity. It was demonstrated that the loss of Fendrr is lethal to mice embryos (Grote et al., 2013; Sauvageau et al. 2013); however, the mechanisms of Fendrr-regulated molecular events still remain to be fully understood.

Polycomb is essential for cardiac development because the conditional deletion of polycomb group component proteins Ezh2 or Eed in cardiomyocytes leads to inappropriate activation of transcriptional factors and cell-cycle regulators, leading to a variable spectrum of congenital and postnatal abnormalities, including septal defects, ventricular hypoplasia and noncompaction, hypertrophic cardiomyopathy, and fibrosis (He et al. 2012; Delgado et al. 2012).

LncRNA Novlnc6

Hundreds of human orthologues of novel lncRNAs modulating two human cardiac disease states, namely dilated cardiomyopathy and aortic stenosis were reported. One of the novel lncRNAs, Novlnc6, is known to regulate cardiac differentiation and maturation through enhancing the expression of BMP10 and NKX2 transcription factors (Ounzain et al. 2015). LncRNA transcriptomic profiling in the human cardiac precursor cell (CPC) revealed 570 lncRNAs that were modulated during cardiac differentiation.

LncRNA H19

Human (GWAS) studies have demonstrated significant associations between the H19 locus and systolic or mean arterial blood pressure (Tragante et al. 2014). High H19 expression has been linked to hyperhomocysteinemia, a risk factor for CAD (Devlin et al. 2005). H19 sponges let-7 family miRNAs, which are believed to have atheroprotective (Satoh et al. 2012) or proatherosclerotic (Qin et al. 2012) roles, are down-regulated in CAD patients (Fichtlscherer et al. 2010).

LncRNA CARMEN

Cardiac mesoderm enhancer-associated non-coding RNA (*CARMEN*), which controls and interacts with PRC2 components, is thus a critical regulator of cardiac cell differentiation and homeostasis (Ounzain et al. 2015). Also, the knockdown of two lncRNAs in mouse C2C12 skeletal myoblasts, that is, n411949 and n413445, involved in the NF-kB pathway by leucine metabolization and Relb modulations, respectively, point towards their role in human cardiac pathology (Matkovich et al. 2014).

LncRNA Kcnq1ot

Transcription of the paternally expressed antisense non-coding RNA Kcnq1ot1 silences some neighboring genes in the embryo. In wild-type mice, it was shown that potassium channel Kv7.1 that encodes transcript KCNQ1OTI expression is regulated by the lncRNA potassium voltage-gated channel KQT-like subfamily, member 1 opposite strand/antisense transcript 1 (Kcnq1ot1). Dysregulation of this lncRNA leads to left ventricular dysfunction as a myocardial infraction complication. The Kcnq1 gene is crucial for heart development and function and is a novel mechanism by which an antisense non-coding RNA affects transcription through regulating chromatin flexibility and access to enhancers (Korostowski et al. 2012).

LncRNA SENCR

LncRNA SENCR (Smooth muscle and endothelial cell enriched migration/differentiation-associated long non-coding RNA) is a human vascular-enriched cytoplasmic lncRNA that is known to be involved in stabilization and differentiation of human (VSMCs) vascular smooth muscle cells (Bell et al. 2014). A novel study reported the regulatory role of SENCR in endothelial development and function: SENCR contributes to the regulation of endothelial differentiation from pluripotent cells, induces proliferation and migration, and controls the angiogenic capacity of human umbilical endothelial cells (HUVECs) (Boulberdaa et al. 2016). Using RNA sequencing, novel lncRNAs in human saphenous vein vascular smooth muscle cells (VSMCs) were discovered, one of which was a smooth-muscle-induced lncRNA (*SMILR*). On stimulation, it is known to act as a VSMC proliferation promoter through the regulation of hyaluronan synthase 2 (HAS2) expression. Thus, it could be a potential target for atherosclerosis therapy (Ballantyne et al. 2016).

LncRNA MALAT1

Metastasis-associated lung adenocarcinoma transcript 1 (*MALAT1*) lncRNA is found mostly in mammalian cells, that is, it is highly expressed in hypoxic endothelial cells. Its function is in vascularization and regulating gene expression during vascular disease development; it also has potential therapeutic benefit (Michalik et al. 2014).

LncRNA GAS5

In humans and rodents, GAS5 was down-regulated during vascular remodeling and hypertension. It also regulates EC and VSMC function through β-catenin signaling *in vivo* and *in vitro* and regulates blood pressure and arterial remodeling (Wang 2016). The diagnostic value of the lncRNA GAS5 for coronary artery disease and exploring the mechanism of the role of GAS5 in CAD has been revealed. GAS5 plays a role as an upstream regulator of the mTOR pathway in human primary coronary artery endothelial cells (HCAECs) and was specifically down-regulated in patients with CAD; it thus may serve as a promising biomarker for CAD (Yin et al. 2017).

LncRNA ANRIL

ANRIL (antisense non-coding RNA in the INK4 locus) expression is associated with atherosclerosis severity, CAD, myocardial infarction, and artery disease. ANRIL leads to an increased rate of cell proliferation, cell adhesion, and a decreased rate of apoptosis, essential mechanisms of atherogenesis. ANRIL binds to polycomb group proteins highly enriched in the proximity of Alu motifs across the genome; the Alu motifs are recruited to be the promoters of target genes during ANRIL over-expression thus regulating the target-genes. The functional importance of Alu motifs was confirmed by deletion, mutagenesis, reversing trans-regulation, and atherogenic cell functions (Holdt et al. 2013).

LncRNA RNCR3

In human and rodent endothelial and vascular smooth muscle cells upon ox-LDL treatment *in vitro*, Retinal non-coding RNA 3 (*RNCR3*) lncRNA expression is significantly up-regulated. It also accelerates apoptotic cell death and deescalates migration. While, *in vivo*, altered expression decreases EC and VSMC proliferation and aggravates hypercholesterolemia and the inflammatory factor release. In ECs, RNCR3 regulates the transcription factor KLF2 by sponging miR-185-5p, which targets KLF2. Thus, due to its atheroprotective role during atherosclerosis, its induced up-regulation potentially represents a therapeutic benefit (Shan et al. 2016).

LncRNA LincRNA-p21

Many lincRNAs play an essential role during p53-dependent transcriptional pathways and are known to be associated with CADs. LincRNA-p21 is one of the lincRNAs that is considered as a potential p53 transcriptional target; it is known to activate apoptosis while binding to hnRNP-K and also in the proper localization of hnRNP-K (Huarte et al. 2010). *In vivo*, it protects against neointimal hyperplasia in rodents and its down-regulation dysregulates many p53 targets, thus leading to the formation of atherosclerotic plaque in CAD patients. *In vitro*, lincRNA-p21 represses cell proliferation rates and induces apoptotic cell death in VSMCs and mouse mononuclear macrophage cells (Wu et al. 2014). Polymorphisms in lincRNA-p21 have been associated with CAD risk as well (Tang et al. 2016). The therapeutic potential of lncRNA-p21 in acute vascular injury is also suggested by its regulatory role in CAD.

LncRNA HAS2-AS1 and SMILR

HAS2 is involved in the process of extracellular matrix remodeling and neointimal formation by hyaluronan synthesis in CAD. Two lncRNAs independently regulate HAS2: HAS2-AS1 (HAS2 antisense RNA 1) and SMILR. HAS2-AS1 was reported to be elevated in atherectomy patient samples in VSMCs by altering chromatin configuration (Vigetti et al., 2014). Smooth Muscle Enriched Long Noncoding RNA (SMILR) up-regulates VSMC proliferation with the involvement of the platelet-derived growth factor and the stimulation of interleukin-1a. Its modulation also is known to have an effect on vascular smooth muscle cell biology and pathology. On stimulation, it is up-regulated in human carotid artery atherosclerotic plaques as compared with adjacent non-stimulated healthy tissues (Ballantyne et al., 2016).

Many LncRNAs are known to have prognostic value and be associated with MI. LncRNA ANRIL, MIAT, and MALAT1 are associated with left ventricular dysfunction as their expression levels change in peripheral blood cells in patients with acute MI (Vausort et al. 2014).

LncRNA MIRT

Two lncRNAs, myocardial infarction-associated transcript 1 and 2 (MIRT1 and MIRT2), regulate the expression of genes, which are involved in reversing left ventricular remodeling and in ejection fraction preservation during myocardial infarction pathogenesis in rodents. Thus, this provides evidence for the therapeutic potential of these two lncRNAs (Zangrando et al. 2014).

LncRNA LIPCAR

Long intergenic non-coding RNA predicting cardiac remodeling (LIPCAR) has a prognostic value in the diagnosis of acute MI. Its levels during heart failure are directly correlated with higher mortality risk in such patients (Kumarswamy et al. 2014) and its levels were elevated in patient plasma samples after acute MI and also in type 2 diabetes mellitus (de Gonzalo-Calvo et al. 2016). This indicates its role as a clinical biomarker.

Other lncRNAs

An in silico approach identified MICRA (myocardial infarction-associated circular RNA) to be the first circRNA biomarker of LV dysfunction (Vausort et al. 2016). Microarray analysis revealed the lncRNA urothelial carcinoma-associated 1 (UCA1) that enhances the process of cardiomyocytic apoptosis leading to the ischemic cardiac injury through enhanced p27 protein levels. Thus, it is considered to be a biomarker in MI (Liu et al. 2015). Another lncRNA that plays a role after heart ischemia is HIF1A-AS2, which is also known as aHIF. HIF1A-AS2 is expressed during cellular response post-ischemic angiogenesis. It acts as a master transcriptional regulator in heart response to ischemia (Zolk et al. 2008). Besides being over-expressed during heart failure, aHIF is under-expressed in the blood of patients after MI (Vausort et al. 2014).

Autophagy is also known to be involved in various cardiovascular diseases and acts as a cardio protectant during ischemia. The lncRNA autophagy-promoting factor (APF) has been identified as the promoter for up-regulating autophagic cell death by targeting ATG7 via miR-188-3p in the MI rodent model (Wang et al. 2015). The cardiac hypertrophy related factor (CHRF) reported as the first lncRNA has been shown to have implications during heart failure and stimulates cardiac hypertrophy. In humans, the over-expression of this lncRNA was observed in heart failure tissue. It is known to regulate cardiac hypertrophy through miR-489, which is an anti-hypertrophic molecule. It inhibits hypertrophy through repressing the Myd88 (target molecule) (Wang et al. 2014).

The cardiac hypertrophy-associated epigenetic regulator (Chaer) is the first lncRNA involved in the hypertrophy of heart muscles in humans and rodents. It directly interacts with polycomb repressor complex 2 (PRC2), which is transient in nature. This interaction is a prerequisite for the epigenetic induction of hypertrophic genes (Wang et al. 2016).

The cardiac hypertrophy-associated transcript (*CHAST*) is a prohypertrophic lncRNA that enhances cardiac hypertrophy. Elevated levels of CHAST have been reported in aortic stenosis patients and are also vital players in *in vivo* and *in vitro* human and mice models (Viereck et al. 2016). Myosin heavy chain-associated RNA transcripts (Mhrt) are potential lncRNAs for the contraction of cardiac muscles and for providing significant cardioprotection during disease conditions by inhibiting myopathy (Han et al. 2014). Mhrt acts as a decoy to inhibit the aberrant expression of pathogenic genes involved in cardiac contractility, thus maintaining cardiac function and providing protection.

In the plasma of CAD patients and controls, microarray analysis revealed many differentially expressed lncRNAs. Coro marker is one of the most highly sensitive, stable, and specific plasma biomarkers for CAD (Yang et al. 2015). Its down-regulation results in the reduced secretion of pro-inflammatory cytokines (IL)-1β, IL-6, and TNF-α in CAD (Cai et al. 2016).

LncAng362 is proximal to miR-221 and miR-222 miRNAs, which regulate the proliferation of rat VSMCs (Liu et al. 2009). The knockdown of lnc-Ang362 down-regulated the expression of miR-221/222 and the proliferation of VSMCs reduction, suggesting that it plays a role in cell growth. Currently, lnc-Ang362 is considered as a therapeutic target for hypertensive patients (Leung et al. 2013).

Future perspectives and conclusions

The discovery of lncRNAs has revolutionized our understanding of disease-regulatory circuits. As more and more lncRNAs are linked to cardiovascular physiology, important questions remained to be addressed before we can fully harness the therapeutic potential of these novel biological modulators. For one, how lncRNAs are targeted to specific genetic loci remains poorly understood. Indeed, the temporal-spatial specificity of lncRNA effects is particularly mystifying given that a large number appear to interact with highly promiscuous factors, such as PRC2 (polycomb repressive complex) (Davidovich et al. 2013). The biochemical and structural basis for why some lncRNAs that resemble mRNAs are retained in the nucleus is also mysterious. Finally, an important bottom line question is, can we leverage the diagnostic or therapeutic potential of lncRNA to advance human health? Although this has been an elusive goal, promising translational efforts are not far behind.

Conflict of interest

The authors declare that they have no conflicts of interest with respect to research, authorship, and publication of this book chapter.

Copyright and permission statement

We confirm that the materials included in this chapter do not violate copyright laws. Where relevant, appropriate permissions have been obtained from the original copyright holder(s). All original sources have been appropriately acknowledged and/or referenced.

Funding

No funding was received for this project.

References

Aird W. C. (2004). Endothelium as an organ system. *Critical care medicine*, *32*(5), S271–S279.

Ballantyne M. D., Pinel K., Dakin R., Vesey A. T., Diver L., Mackenzie R., . . . & Joshi N. (2016). Smooth muscle enriched long noncoding RNA (SMILR) regulates cell proliferation. *Circulation*, *133*(21), 2050–65.

Bazzoni G., & Dejana E. (2004). Endothelial cell-to-cell junctions: molecular organization and role in vascular homeostasis. *Physiological reviews*, *84*(3), 869–901.

Bell R. D., Long X., Lin M., Bergmann J. H., Nanda V., Cowan S. L., . . . & Miano J. M. (2014). Identification and initial functional characterization of a human vascular cell–enriched long noncoding RNA. *Arteriosclerosis, thrombosis, and vascular biology*, *34*(6), 1249–59.

Boulberdaa M., Scott E., Ballantyne M., Garcia R., Descamps B., Angelini G. D., . . . & Miano J. M. (2016). A role for the long noncoding RNA SENCR in commitment and function of endothelial cells. *Molecular Therapy*, *24*(5), 978–90.

Cai Y., Yang Y., Chen X., Wu G., Zhang X., Liu Y., . . . & Jose P. A. (2016). Circulating “lncRNA OTTHUMT00000387022” from monocytes as a novel biomarker for coronary artery disease. *Cardiovascular research*, *112*(3), 714–24.

Cech T. R., & Steitz J. A. (2014). The noncoding RNA revolution—trashing old rules to forge new ones. *Cell*, 157(1), 77–94.

ENCODE Project Consortium. (2004). The ENCODE (Encyclopedia of DNA elements) project. *Science*, *306*(5696), 636–40.

Danesh J., Wheeler J. G., Hirschfield G. M., Eda S., Eiriksdottir G., Rumley A., . . . & Gudnason V. (2004). C-reactive protein and other circulating markers of inflammation in the prediction of coronary heart disease. *New England journal of medicine*, *350*(14), 1387–97.

Davidovich C., Zheng L., Goodrich K. J., & Cech T. R. (2013). Promiscuous RNA binding by Polycomb repressive complex 2. *Nature structural & molecular biology*, *20*(11), 1250.

de Gonzalo-Calvo D., Kenneweg F., Bang C., Toro R., Van Der Meer R. W., Rijzewijk L. J., . . . & Thum T. (2016). Circulating long-non coding RNAs as biomarkers of left ventricular diastolic function and remodelling in patients with well-controlled type 2 diabetes. *Scientific reports*, *6*, 37354.

Delgado-Olguín P., Huang Y., Li X., Christodoulou D., Seidman C. E., Seidman J. G., . . . & Bruneau, B. G. (2012). Epigenetic repression of cardiac progenitor gene expression by Ezh2 is required for postnatal cardiac homeostasis. *Nature genetics*, *44*(3), 343.

Devlin A. M., Bottiglieri T., Domann F. E., & Lentz, S. R. (2005). Tissue-specific changes in H19 methylation and expression in mice with hyperhomocysteinemia. *Journal of biological chemistry*, *280*(27), 25506–511.

Di Salvo T. G., Guo Y., Su Y. R., Clark T., Brittain E., Absi T., . . . & Hemnes A. (2015). Right ventricular long noncoding RNA expression in human heart failure. *Pulmonary circulation*, *5*(1), 135–61.

Esteller M. (2011). Non-coding RNAs in human disease. *Nature reviews genetics*, *12*(12), 861.

Fichtlscherer S., De Rosa S., Fox H., Schwietz T., Fischer A., Liebetrau C., . . . & Bonauer A. (2010). Circulating microRNAs in patients with coronary artery disease. *Circulation research*, *107*(5), 677–84.

Grote P., Wittler L., Hendrix D., Koch F., Währisch S., Beisaw A., . . . & Herrmann B. G. (2013). The tissue-specific lncRNA Fendrr is an essential regulator of heart and body wall development in the mouse. *Developmental cell*, *24*(2), 206–14.

Gu S., Li G., Zhang X., Yan J., Gao J., An X., . . . & Su P. (2015). Aberrant expression of long noncoding RNAs in chronic thromboembolic pulmonary hypertension. *Molecular medicine reports*, *11*(4), 2631–43.

Han P., Li W., Lin C. H., Yang J., Shang C., Nurnberg S. T., . . . & Xiong Y. (2014). A long noncoding RNA protects the heart from pathological hypertrophy. *Nature*, *514*(7520), 102.

He A., Ma Q., Cao J., Von Gise A., Zhou P., Xie H., . . . & Daley G. Q. (2012). Polycomb repressive complex 2 regulates normal development of the mouse heart. *Circulation research*, *110*(3), 406–15.

Holdt L. M., Hoffmann S., Sass K., Langenberger D., Scholz M., Krohn K., . . . & Gielen S. (2013). Alu elements in ANRIL non-coding RNA at chromosome 9p21 modulate atherogenic cell functions through trans-regulation of gene networks. *PLoS genetics*, *9*(7), e1003588.

Huarte M. (2015). The emerging role of lncRNAs in cancer. *Nature medicine*, *21*(11), 1253.

Huarte M., Guttman M., Feldser D., Garber M., Koziol M. J., Kenzelmann-Broz D., . . . & Attardi, L. D. (2010). A large intergenic noncoding RNA induced by p53 mediates global gene repression in the p53 response. *Cell*, *142*(3), 409–19.

Sauvageau M., Goff L. A., Lodato S., Bonev B., Groff A. F., Gerhardinger C., . . . & Liapis, S. C. (2013). Multiple knockout mouse models reveal lincRNAs are required for life and brain development. *elife*, *2*, e01749.

Jiang W., Liu Y., Liu R., Zhang K., & Zhang Y. (2015). The lncRNA DEANR1 facilitates human endoderm differentiation by activating FOXA2 expression. *Cell reports*, *11*(1), 137–48.

Korostowski L., Sedlak N., & Engel N. (2012). The Kcnq1ot1 long non-coding RNA affects chromatin conformation and expression of Kcnq1, but does not regulate its imprinting in the developing heart. *PLoS genetics*, *8*(9), e1002956.

Kumarswamy R., Bauters C., Volkmann I., Maury F., Fetisch J., Holzmann A., . . . & Thum T. (2014). Circulating long noncoding RNA, LIPCAR, predicts survival in patients with heart failure. *Circulation research*, *114*(10), 1569–75.

Leung A., Trac C., Jin W., Lanting L., Akbany A., Sætrom P., . . . & Natarajan R. (2013). Novel long noncoding RNAs are regulated by angiotensin II in vascular smooth muscle cells. *Circulation research*, *113*(3), 266–78.

Liu X., Cheng Y., Zhang S., Lin Y., Yang J., & Zhang C. (2009). A necessary role of miR-221 and miR-222 in vascular smooth muscle cell proliferation and neointimal hyperplasia. *Circulation research*, *104*(4), 476–87.

Liu Y., Zhou D., Li G., Ming X., feng Tu Y., Tian J., . . . & Yu B. (2015). Long non-coding RNA-UCA1 contributes to cardiomyocyte apoptosis by suppression of p27 expression. *Cellular physiology and biochemistry*, *35*(5), 1986–98.

Loewer S., Cabili M. N., Guttman M., Loh Y. H., Thomas K., Park I. H., . . . & Manos P. D. (2010). Large intergenic non-coding RNA-RoR modulates reprogramming of human induced pluripotent stem cells. *Nature genetics*, *42*(12), 1113.

Lorenzen J. M., & Thum T. (2016). Long noncoding RNAs in kidney and cardiovascular diseases. *Nature reviews nephrology*, *12*(6), 360.

Matkovich S. J., Edwards J. R., Grossenheider T. C., de Guzman Strong C., & Dorn G. W. (2014). Epigenetic coordination of embryonic heart transcription by dynamically regulated long noncoding RNAs. *Proceedings of the National Academy of Sciences*, *111*(33), 12264–69.

Michalik K. M., You X., Manavski Y., Doddaballapur A., Zörnig M., Braun T., . . . & Boon R. A. (2014). Long noncoding RNA MALAT1 regulates endothelial cell function and vessel growth. *Circulation research*, *114*(9), 1389–97.

O'Donnell K. A., Wentzel E. A., Zeller K. I., Dang C. V., & Mendell J. T. (2005). c-Myc-regulated microRNAs modulate E2F1 expression. *Nature*, *435*(7043), 839.

Ounzain S., Micheletti R., Arnan C., Plaisance I., Cecchi D., Schroen B., . . . & Bussotti G. (2015). CARMEN, a human super enhancer-associated long noncoding RNA controlling cardiac specification, differentiation and homeostasis. *Journal of molecular and cellular cardiology*, *89*, 98–112.

Ounzain S., Micheletti R., Beckmann T., Schroen B., Alexanian M., Pezzuto I., . . . & Dauvillier J. (2014). Genome-wide profiling of the cardiac transcriptome after myocardial infarction identifies novel heart-specific long non-coding RNAs. *European heart journal*, *36*(6), 353–68.

Qin B., Xiao B., Liang D., Li Y., Jiang T., & Yang H. (2012). MicroRNA let-7c inhibits Bcl-xl expression and regulates ox-LDL-induced endothelial apoptosis. *BMB reports*, *45*(8), 464–69.

Rinn J. L., & Chang H. Y. (2012). Genome regulation by long noncoding RNAs. *Annual review of biochemistry*, *81*, 145–66.

Sahu A., Singhal U., & Chinnaiyan A. M. (2015). Long noncoding RNAs in cancer: from function to translation. *Trends in cancer*, *1*(2), 93–109.

Satoh M., Tabuchi T., Minami Y., Takahashi Y., Itoh T., & Nakamura M. (2012). Expression of let-7i is associated with Toll-like receptor 4 signal in coronary artery disease: effect of statins on let-7i and Toll-like receptor 4 signal. *Immunobiology*, *217*(5), 533–39.

Shan K., Jiang Q., Wang X. Q., Yang H., Yao M. D., Liu C., . . . & Yong J. (2016). Role of long non-coding RNA-RNCR3 in atherosclerosis-related vascular dysfunction. *Cell death & disease*, *7*(6), e2248.

Tang S. S., Cheng J., Cai M. Y., Yang X. L., Liu X. G., Zheng B. Y., & Xiong X. D. (2016). Association of lincRNA-p21 haplotype with coronary artery disease in a Chinese Han population. *Disease markers*, *2016*.

Thum T., & Condorelli G. (2015). Long noncoding RNAs and microRNA in cardiovascular pathophysiology. *Circulation research*, *116*(4), 751–62.

Tragante V., Barnes M. R., Ganesh S. K., Lanktree M. B., Guo W., Franceschini N., . . . & Karczewski K. J. (2014). Gene-centric meta-analysis in 87,736 individuals of European ancestry identifies multiple blood-pressure-related loci. *American journal of human genetics*, *94*(3), 349–60.

Uchida S., & Dimmeler S. (2015). Long noncoding RNAs in cardiovascular diseases. *Circulation research*, 116(4), 737–50.

van Rooij E. (2011). The art of microRNA research. *Circulation research*, *108*(2), 219–34.

Vausort M., Salgado-Somoza A., Zhang L., Leszek P., Scholz M., Teren A., . . . & Devaux Y. (2016). Myocardial infarction-associated circular RNA predicting left ventricular dysfunction. *Journal of the American College of Cardiology*, *68*(11), 1247–48.

Vausort M., Wagner D. R., & Devaux Y. (2014). Long noncoding RNAs in patients with acute myocardial infarction. *Circulation research*, *115*(7), 668–77.

Viereck J., Kumarswamy R., Foinquinos A., Xiao K., Avramopoulos P., Kunz M., . . . & Just A. (2016). Long noncoding RNA Chast promotes cardiac remodeling. *Science translational medicine*, *8*(326), 326ra22–326ra22.

Vigetti D., Deleonibus S., Moretto P., Bowen T., Fischer J. W., Grandoch M., . . . & Karousou E. (2014). Natural antisense transcript for hyaluronan synthase 2 (HAS2-AS1) induces transcription of HAS2 via protein O-GlcNAcylation. *Journal of biological chemistry*, *289*(42), 28816–26.

Wallez Y., & Huber P. (2008). Endothelial adherens and tight junctions in vascular homeostasis, inflammation and angiogenesis. *Biochimica et biophysica acta (BBA)—biomembranes*, *1778*(3), 794–809.

Wang K., Liu C. Y., Zhou L. Y., Wang J. X., Wang M., Zhao B., . . . & Feng C. (2015). APF lncRNA regulates autophagy and myocardial infarction by targeting miR-188-3p. *Nature communications*, *6*, 6779.

Wang K., Liu F., Zhou L. Y., Long B., Yuan S. M., Wang Y., . . . & Li P. F. (2014). The long noncoding RNA CHRF regulates cardiac hypertrophy by targeting miR-489. *Circulation research*, *114*(9), 1377–88.

Wang X., Yan C., Xu X., Dong L., Su H., Hu Y., . . . & Ying K. (2016). Long noncoding RNA expression profiles of hypoxic pulmonary hypertension rat model. *Gene*, *579*(1), 23–28.

Wang Y. N. Z., Shan K., Yao M. D., Yao J., Wang J. J., Li X., . . . & Yan B. (2016). Long noncoding RNA-GAS5: a novel regulator of hypertension-induced vascular remodeling. *Hypertension*, *68*(3), 736–48.

Wang Z., Zhang X. J., Ji Y. X., Zhang P., Deng K. Q., Gong J., . . . & Gao C. (2016). The long noncoding RNA Chaer defines an epigenetic checkpoint in cardiac hypertrophy. *Nature medicine*, *22*(10), 1131.

Ward M., McEwan C., Mills J. D., & Janitz M. (2015). Conservation and tissue-specific transcription patterns of long noncoding RNAs. *Journal of human transcriptome*, *1*(1), 2–9.

West N. E., Qian H., Guzik T. J., Black E., Cai S., George S. E., & Channon K. M. (2001). Nitric oxide synthase (nNOS) gene transfer modifies venous bypass graft remodeling: effects on vascular smooth muscle cell differentiation and superoxide production. *Circulation*, *104*(13), 1526–32.

Wu G., Cai J., Han Y., Chen J., Huang Z. P., Chen C., . . . & Xu Z. (2014). LincRNA-p21 regulates neointima formation, vascular smooth muscle cell proliferation, apoptosis, and atherosclerosis by enhancing p53 activity. *Circulation*, *130*(17), 1452–65.

Xue Z., Hennelly S., Doyle B., Gulati A. A., Novikova I. V., Sanbonmatsu K. Y., & Boyer L. A. (2016). A G-rich motif in the lncRNA Braveheart interacts with a zinc-finger transcription factor to specify the cardiovascular lineage. *Molecular cell*, *64*(1), 37–50.

Yang Y., Cai Y., Wu G., Chen X., Liu Y., Wang X., . . . & Zhou L. (2015). Plasma long non-coding RNA, CoroMarker, a novel biomarker for diagnosis of coronary artery disease. *Clinical Science*, *129*(8), 675–85.

Yin Q., Wu A., & Liu M. (2017). Plasma long non-coding RNA (lncRNA) GAS5 is a new biomarker for coronary artery disease. *Medical science monitor*, *23*, 6042.

Zangrando J., Zhang L., Vausort M., Maskali F., Marie P. Y., Wagner D. R., & Devaux Y. (2014). Identification of candidate long non-coding RNAs in response to myocardial infarction. *BMC genomics*, *15*(1), 460.

Zolk O., Solbach T. F., Eschenhagen T., Weidemann A., & Fromm M. F. (2008). Activation of negative regulators of the hypoxia-inducible factor (HIF) pathway in human end-stage heart failure. *Biochemical and biophysical research communications*, *376*(2), 315–20.

CHAPTER 9

REGULATORY MECHANISMS OF LONG NON-CODING RNAS IN HOST–PATHOGEN INTERACTIONS

SYED MUDASIR AHMAD[1*]

[1] Division of Animal Biotechnology, Faculty of Veterinary Sciences and Animal Husbandry, Shuhama, SKUAST-K-190006, India
*Corresponding author
Tel: 9419514131; E-mail: mudasirbio@gmail.com

Abstract

Non-coding RNAs according to their size are divided into small ncRNAs (<200nt) including miRNA, siRNA, piRNA, and long ncRNA (almost 200nt). Though the functions of most lncRNAs are unknown, it has been shown that many regulate various cellular activities. Recent studies suggest that lncRNAs play critical roles during innate immune responses in transcriptionally regulating gene expression. LncRNA's regulation of inflammatory mediators or cytokines is still poorly understood, but it has been found that lncRNAs can contribute to gene regulatory networks governing host–pathogen interactions. Some lncRNAs, among them, are recognized as significant host response regulators for pathogen contagion, as described in this book chapter.

Keywords: LncRNAs, Host Pathogen interactions, Infection, Macrophage, TLR4

It has become increasingly evident in recent years that non-coding RNAs (ncRNAs) play an important role in diverse biological processes. Non-coding RNAs can be roughly divided into two main categories according to their size: small ncRNAs (<200nt) including miRNA, siRNA, piRNA, and long ncRNA (almost 200nt). Non-coding genome members include miRNA, a major class of well-characterized, conserved, and endogenous small interfering RNA regulating gene expression (Huntzinger and Izaurralde 2011). Micro RNA controls the silencing of post-transcription genes by regulating the translation and RNA stability (Sharbati et al. 2011). Small non-coding RNAs play a major part in the host response to bacterial agents. However, the function of the long non-coding RNA (lncRNA) remains unknown in this context (Pawar et al. 2016). Though the functions of most lncRNAs are unknown, it has been shown that many regulate various cellular activities. Recent studies suggest that lncRNAs may also play critical roles during innate immune responses in transcriptionally regulating gene expression. Dysfunctional performance of lncRNAs has been found to have a significant effect on polygenic diseases caused by cancers of different organs (Li et al. 2013) to non-cancerous diseases, such as Alzheimer's disease (Faghihi et al. 2008). Multiple theories have emerged recently about the role of lncRNA in infectious diseases and some studies find that several lncRNAs may be involved in controlling immune responses (Scaria et al. 2012). LncRNA's regulation of inflammatory mediators or cytokines is still poorly understood, but it has been found that lncRNAs can contribute to gene regulatory networks governing host–pathogen interactions (Geng and Tan 2016).

Cellular response to infection relies on well-defined molecular interaction networks based on gene expression regulation and protein function. Current research has discovered a considerable number of lncRNAs associated with disease states under infection. Some lncRNAs, among them, are recognized as significant host response regulators for pathogen contagion. However, given that the roles of lncRNAs are spatio-temporally controlled, analysis has rejected time-dependent influence or localization-dependent function of lncRNAs. Therefore, it is conceivable that unknown lncRNAs are present whose levels of expression do not vary in infection, but depend upon altering their patterns of localization. Furthermore, NGS analysis using bulk and mixed samples neglect the host cell heterogeneity or the dissimilarity of the infection stages, which leads to difficulties in recognizing those functional lncRNAs whose expression intensities fluctuate between individual host cells. The recent single cell RNA-sequencing approach (scRNA-seq) for infected cells is more instructive (Cristinelli and Ciuffi 2018); however, the ability to acquire satisfactorily read depths of infrequently expressed RNAs like pathogen-derived RNAs through scRNA-seq is inadequate.

Kawai and Akira (2010) found that toll-like receptors (TLRs) have a major role in recognition of pathogens. Specific signaling pathways in the synthesis of inflammatory mediators or cytokines are activated following ligand recognition by these receptors. This involves the regulation of transcription or post-transcription genes via transcription factors such as API and NF-kB and miRNAs (Kawai and Akira 2010). The macrophage is a key component of host defense mechanisms against pathogens. The first line of defense against invading pathogens is represented by phagocytic cells like macrophages and polymorphonuclear neutrophilic granulocytes (PMNs). Stimulation of human primary monocytes by lipopolysaccharide (LPS), which is a major component of gram-negative bacteria's outer membrane, stimulates bidirectional transcription of various lncRNAs and enhancer RNAs. The enhancer RNA IL1β-eRNA surrounds the locus IL1β and the transcript IL1β-RBT46 is transcribed bidirectionally. The enhancer RNAs IL1β-eRNA and IL1β-RBT46 both regulate transcription of CXCL8 and IL1β induced by lipopolysaccharide (Ilott et al. 2014). The long-intervening ncRNA (lincRNA)-Cox2 is an essential regulator stimulated during LPS treatment in CD11c+ bone-marrow-derived dendritic cells (Guttman et al. 2009) and in macrophages emerging from mouse bone marrow, the lincRNA-Cox2 attachment with heterogeneous nuclear ribonucleoprotein A/B and A2/B1 stimulate and inhibit genes connected with inflammatory reactions (Carpenter et al. 2013). LincRNA-Cox2 functions have been recently established using lincRNA-Cox2 KO mice and are known to control its neighboring gene Ptgs2 in both cis and trans (Elling et al. 2018). LPS prompts the transitory production of ncRNA in HD11 cells. The transcription from the upstream region of the lysozyme gene marks triggering the expression of lysozyme (Lefevre et al. 2008).

In human hepatocarcinoma-derived HuH7 cells, both lncISG15 and lncBST2/BISPR were identified upon treatment with IFNα2 as well as by infection with hepatitis C virus. Induced by an interferon-dependent pathway, these lncRNAs are expressed near interferon-stimulated gene loci. The RNAi-mediated down-regulation of lncBST2/BISPR decreases the expression of BST2 and indicates its regulatory function in the expression of its adjacent genes (Barriocanal et al. 2014).

Some lncRNAs control the gene expression of immune sensitive genes by networking with enzymes linked to histone modification. For example, in both humans and mice, the lncRNA gene NeST is situated neighboring to the interferon γ (IFN-γ) translating gene. In both genomes, NeST RNA is coded in the opposite DNA strand to that coding for IFN-π, and both the genes are transcribed by converging (Vigneau et al. 2001). NeST binds to a part of the H3 lysine 4-methyltransferase complexes called WDR5, thereby increasing the expression of IFN-γ locus (Gomez et al. 2013). In CD8 + T cells, tuberculosis-induced lncRNA-CD244 recruits zeste homolog 2 (EZH2) enhancer and polycomb protein, thus augmenting the trimethylation of H3K27 in promoter regions of the gene IFN-π / TNF-α, and thereby reducing these gene expressions (Wang et al. 2015). On infection with a variety of viruses such as IAV, Muscovy duck reovirus, Sendai virus, and herpes simplex virus (HSV) on human alveolar epithelial cells, the expression of the lncRNA NRAV is inhibited while as on the over-expression of lncRNA NRAV, interferon stimulated genes (ISGs) like MxA and IFITM3 are repressed at their transcription start sites with a decrease of H3K4me3. To summarize, NRAV functions as a negative regulator of ISGs (Ouyang et al. 2014). The treatment of human hepatocytes by poly (I: C), which imitates viral double-stranded RNA, identified lncRNA-LUARIS via microarray analysis. LUARIS, being an antisense transcript of the HECW1 protein-coding gene, is known to be inhibited by interferon signaling type I. The SiRNA knockdown of LUARIS leads to the reduced expression of ISGs as it interacts directly with nuclear heterogeneous ribonucleoprotein U (hnRNPU) followed by the activation of transcription factor 2 (ATF2) (Nishitsuji et al. 2016).

NEAT1 is an architectural long non-coding RNA that produces paraspeckles and contributes an antiviral effect. Up-regulation of NEAT1 occurs in HeLa cells upon infection with the influenza virus and HSV, and in non-infected cells causes the binding of the splicing factor proline- and glutamine-rich SFPQ to the IL8 promoter region. However, after poly (I:C) treatment, the induction of NEAT1 causes SFPQ sequestration into paraspeckles and thus the expression of antiviral cytokines such as IL-8 (Imamura et al. 2014). Hantaan virus infection on human umbilical vein endothelial cells cause NEAT1 to sequester SFPQ to paraspeckles and thus the activation of the RIG-I-IRF7 pathway by inducing the transcription of DDX60 and RIG-I followed by interferon-β production (Ma et al. 2017). In HeLa cells, NEAT1 interacts with a transcription inhibitor, HEXIM1 and this complex NEAT1-HEXIM1 via the cGAS-STING pathway controls the immune response on infection with Kaposi's sarcoma-associated herpes virus (Morchikh et al. 2017).

Another method for the functioning of lncRNA in the protection of the host is to prevent pathogen replication. For example, the lncRNA NRON suppresses replication of human immunodeficiency virus type 1 (HIV-1) by constraining the transcription nuclear factor of activated T cells (NFAT) known to augment HIV-1 viral replication in Jurkat cells (Imam et al. 2015). The lncRNA NRON also degrades a viral transcription activator protein called Tat in TZM-bl cells by cooperating with the ubiquitin ligases PSMD11, CUL4B, and HUWE1 (Li et al. 2016). NEAT1 also suppresses HIV-1 growth through exporting HIV-1 mRNAs to the cytoplasm via a rev-dependent pathway (Malim et al. 1989). While NEAT1 down-regulation, results in localization of HIV-1 mRNAs (non-spliced form) in the cytoplasm of HeLa cells (Zhang et al. 2013). NEAT1 and paraspeckle protein components are also known to interact directly with viral genomes and via the transcription factor STAT3 regulate both viral replication and gene expression (Wang et al. 2017). Similarly, lncRNA GAS5, a growth-arrest specific transcript (Scheider et al. 1988) in cancer (Pickard et al. 2013), acts as a decoy of the hepatitis C virus NS3 protein and thus counteracts virus reproduction in Huh7 cells by preventing viral replication and assemblage (Qian et al. 2016).

Interestingly, a few studies have highlighted the role that lncRNAs play in regulating autophagy in cancer. For example, decreased expression of the lncRNA MEG3 has been shown to result in activation of autophagy in cancer cells in the bladder (Ying et al. 2013). However, the molecular details of the mode of action are not comprehended. In a recent study, it has been shown that mycobacterial infection and human macrophage stimulation by IFN-ÿ lead to MEG3-mediated induction of autophagy in intracellular bacteria (Sharbati et al. 2011)

In the end, developing novel tools and technologies to detect changes in RNA localization and overcome read-depth limitations will ascertain new characteristics of lncRNA-mediated reactions to pathogen infectivity. The characterization of lncRNA genes and their modes of action will enable them to be used for diagnosis, progression monitoring, and targeted therapy in various infectious diseases.

Conflict of interest

The author declares that he has no conflicts of interest with respect to research, authorship, and publication of this book chapter.

Copyright and permission statement

We confirm that the materials included in this chapter do not violate copyright laws. Where relevant, appropriate permissions have been obtained from the original copyright holder(s). All original sources have been appropriately acknowledged and/or referenced.

Funding

No funding was received for this project.

References

Barriocanal, M., Carnero, E., Segura, V., & Fortes, P. (2015). Long non-coding RNA BST2/BISPR is induced by IFN and regulates the expression of the antiviral factor tetherin. *Frontiers in immunology*, *5*, 655.

Carpenter, S., Aiello, D., Atianand, M. K., Ricci, E. P., Gandhi, P., Hall, L. L., . . . & O'Neill, L. A. (2013). A long noncoding RNA mediates both activation and repression of immune response genes. *Science*, *341*(6147), 789–92.

Cristinelli, S., & Ciuffi, A. (2018). The use of single-cell RNA-Seq to understand virus–host interactions. *Current opinion in virology*, *29*, 39–50.

Elling, R., Robinson, E. K., Shapleigh, B., Liapis, S. C., Covarrubias, S., Katzman, S., . . . & Chan, J. (2018). Genetic models reveal cis and trans immune-regulatory activities for lincRNA-Cox2. *Cell reports*, *25*(6), 1511–24.

Faghihi, M. A., Modarresi, F., Khalil, A. M., Wood, D. E., Sahagan, B. G., Morgan, T. E., . . . & Wahlestedt, C. (2008). Expression of a noncoding RNA is elevated in Alzheimer's disease and drives rapid feed-forward regulation of β-secretase. *Nature medicine*, *14*(7), 723–30.

Geng, H., & Tan, X. D. (2016). Functional diversity of long non-coding RNAs in immune regulation. *Genes & diseases*, *3*(1), 72–81.

Gomez, J. A., Wapinski, O. L., Yang, Y. W., Bureau, J. F., Gopinath, S., Monack, D. M., . . . & Kirkegaard, K. (2013). The NeST long ncRNA controls microbial susceptibility and epigenetic activation of the interferon-γ locus. *Cell*, *152*(4), 743–54.

Guttman, M., Amit, I., Garber, M., French, C., Lin, M. F., Feldser, D., . . . & Cabili, M. N. (2009). Chromatin signature reveals over a thousand highly conserved large non-coding RNAs in mammals. Nature, *458*(7235), 223–27.

Huntzinger, E., & Izaurralde, E. (2011). Gene silencing by microRNAs: contributions of translational repression and mRNA decay. *Nature reviews genetics*, *12*(2), 99–110.

Ilott, N. E., Heward, J. A., Roux, B., Tsitsiou, E., Fenwick, P. S., Lenzi, L., . . . & Donnelly, L. E. (2014). Long non-coding RNAs and enhancer RNAs regulate the lipopolysaccharide-induced inflammatory response in human monocytes. *Nature communications*, *5*(1), 1–14.

Imam, H., Bano, A. S., Patel, P., Holla, P., & Jameel, S. (2015). The lncRNA NRON modulates HIV-1 replication in a NFAT-dependent manner and is differentially regulated by early and late viral proteins. *Scientific reports*, *5*, 8639.

Imamura, K., Imamachi, N., Akizuki, G., Kumakura, M., Kawaguchi, A., Nagata, K., . . . & Kai, C. (2014). Long noncoding RNA NEAT1-dependent SFPQ relocation from promoter region to paraspeckle mediates IL8 expression upon immune stimuli. *Molecular cell*, *53*(3), 393–406.

Kawai, T., & Akira, S. (2010). The role of pattern-recognition receptors in innate immunity: update on Toll-like receptors. *Nature immunology*, *11*(5), 373.

Lefevre, P., Witham, J., Lacroix, C. E., Cockerill, P. N., & Bonifer, C. (2008). The LPS-induced transcriptional upregulation of the chicken lysozyme locus involves CTCF eviction and noncoding RNA transcription. *Molecular cell*, *32*(1), 129–39.

Li, J., Chen, C., Ma, X., Geng, G., Liu, B., Zhang, Y., . . . & Cai, W. (2016). Long noncoding RNA NRON contributes to HIV-1 latency by specifically inducing tat protein degradation. *Nature communications*, *7*(1), 1–10.

Li, J., Xuan, Z., & Liu, C. (2013). Long non-coding RNAs and complex human diseases. *International journal of molecular sciences*, *14*(9), 18790–808.

Ma, H., Han, P., Ye, W., Chen, H., Zheng, X., Cheng, L., . . . & Zhang, F. (2017). The long noncoding RNA NEAT1 exerts antihantaviral effects by acting as positive feedback for RIG-I signaling. *Journal of virology*, *91*(9), e02250-16.

Malim, M. H., Hauber, J., Le, S. Y., Maizel, J. V., & Cullen, B. R. (1989). The HIV-1 rev trans-activator acts through a structured target sequence to activate nuclear export of unspliced viral mRNA. *Nature*, *338*(6212), 254–57.

Morchikh, M., Cribier, A., Raffel, R., Amraoui, S., Cau, J., Severac, D., . . . & Benkirane, M. (2017). HEXIM1 and NEAT1 long non-coding RNA form a multi-subunit complex that regulates DNA-mediated innate immune response. *Molecular cell*, *67*(3), 387–99.

Nishitsuji, H., Ujino, S., Yoshio, S., Sugiyama, M., Mizokami, M., Kanto, T., & Shimotohno, K. (2016). Long noncoding RNA# 32 contributes to antiviral responses by controlling interferon-stimulated gene expression. *Proceedings of the National Academy of Sciences*, *113*(37), 10388–93.

Ouyang, J., Zhu, X., Chen, Y., Wei, H., Chen, Q., Chi, X., . . . & Wang, G. (2014). NRAV, a long noncoding RNA, modulates antiviral responses through suppression of interferon-stimulated gene transcription. *Cell host & microbe*, *16*(5), 616–26.

Pawar, K., Hanisch, C., Vera, S. E. P., Einspanier, R., & Sharbati, S. (2016). Down regulated lncRNA MEG3 eliminates mycobacteria in macrophages via autophagy. *Scientific reports*, *6*(1), 1–13.

Pickard, M. R., Mourtada-Maarabouni, M., & Williams, G. T. (2013). Long non-coding RNA GAS5 regulates apoptosis in prostate cancer cell lines. *Biochimica et biophysica acta (BBA)—molecular basis of disease*, *1832*(10), 1613–23.

Qian, X., Xu, C., Zhao, P., & Qi, Z. (2016). Long non-coding RNA GAS5 inhibited hepatitis C virus replication by binding viral NS3 protein. *Virology*, *492*, 155–65.

Scaria, V., & Pasha, A. (2013). Long non-coding RNAs in infection biology. *Frontiers in genetics*, *3*, 308.

Schneider, C., King, R. M., & Philipson, L. (1988). Genes specifically expressed at growth arrest of mammalian cells. *Cell*, *54*(6), 787–93.

Sharbati, J., Lewin, A., Kutz-Lohroff, B., Kamal, E., Einspanier, R., & Sharbati, S. (2011). Integrated microRNA-mRNA-analysis of human monocyte derived macrophages upon Mycobacterium avium subsp. hominissuis infection. *PloS one*, *6*(5).

Vigneau, S., Levillayer, F., Crespeau, H., Cattolico, L., Caudron, B., Bihl, F., . . . & Bureau, J. F. (2001). Homology between a 173-kb region from mouse chromosome 10, telomeric to the Ifng locus, and human chromosome 12q15. *Genomics*, *78*(3), 206–13.

Wang, Y., Zhong, H., Xie, X., Chen, C. Y., Huang, D., Shen, L., . . . & Zeng, G. (2015). Long noncoding RNA derived from CD244 signaling epigenetically controls CD8+ T-cell immune responses in tuberculosis infection. *Proceedings of the National Academy of Sciences*, *112*(29), E3883–E3892.

Wang, Z., Fan, P., Zhao, Y., Zhang, S., Lu, J., Xie, W., . . . & Zhang, Y. (2017). NEAT1 modulates herpes simplex virus-1 replication by regulating viral gene transcription. *Cellular and molecular life sciences*, *74*(6), 1117–31.

Ying, L., Huang, Y., Chen, H., Wang, Y., Xia, L., Chen, Y., . . . & Qiu, F. (2013). Downregulated MEG3 activates autophagy and increases cell proliferation in bladder cancer. *Molecular biosystems*, *9*(3), 407–11.

Zhang, Q., Chen, C. Y., Yedavalli, V. S., & Jeang, K. T. (2013). NEAT1 long noncoding RNA and paraspeckle bodies modulate HIV-1 posttranscriptional expression. MBio 4, e00596–00512.